SAXON MATH™ 2

Guide to Differentiated Instruction

SAXON®

An Imprint of HMH
Supplemental Publishers

www.SaxonPublishers.com
1-800-531-5015

Printed in the U.S.A.

ISBN 978-1-602-77447-6

10 2026

4500930133 ^ B C D E F G

Introduction

While teaching the lessons in the *Saxon Math* program, the teacher may determine that additional instruction is needed for some students, to help them understand the concepts being presented. This may include reviewing vocabulary and presenting the lesson content in simpler terms.

The activities included in this booklet have been developed to complement the lessons in order to assist the teacher in addressing the differing needs of students, especially those requiring help for the following reasons:

- **English Learners:** students for whom English is not the primary language.
- **Inclusion Students:** students for whom concepts may be more difficult to grasp or who may have disabilities that require additional attention.
- **At-Risk Students:** students who have the ability to understand the lesson content as presented, but for various reasons, such as illness or other issues, have fallen behind in understanding.
- **Advanced Learners:** students who have easily grasped the lesson content and may need to be challenged to promote continued growth and enthusiasm for learning.

Tip

The Teacher's Resource Handbook for *Saxon Math K–4* is another excellent source of tips for teachers on meeting the needs of all children in the classroom.

This Guide includes an English Learners activity and an At-Risk Students activity for every lesson in the program. A few lessons will have three activities of the above four, omitting either the Inclusion Students activity or the Advanced Learners activity.

Often a teacher will not know that a student is struggling with a certain concept until after an assessment or some time has passed after the concept has been taught. To assist the child in learning the concept, it is best to go back to the lesson where the content was first taught and help the child do the activities written for that lesson.

Activities in this Guide are intended to be completed with the child after teaching the lesson while the rest of the class is engaged in other activities. Some activities require additional materials, which are usually easily accessible and often have been used during the lesson. These materials are listed with each activity. Common classroom materials such as paper and pencil are not listed as materials needed.

English Learners

The English Learners activities focus on language acquisition through hands-on activities and models. The mathematical ability of an English learner is probably the same as other children in the classroom, but may be hampered by a language barrier.

English Learners

materials: container with 20 color tiles

- **reasonable:** based on good sense (sensible); explainable; thought out
- Explain that when people make decisions based on available information, they are using reasoning; ask the child if he or she would wear a bathing suit in cold weather and to explain why or why not.
- Continue that in the same way, when people make estimates based on available information, they use reasoning; they can judge whether something makes sense or not and when something makes sense, it is reasonable.
- Ask which is a better estimate of the number of color tiles in the container, 20 or 200, and ask why it is a reasonable estimate.

To assist the child in his or her understanding of English terms used, specific words and phrases are selected from the lesson, generally from those listed in the Mathematical Language box on the first page of each lesson. In lessons where the word selected is from a student worksheet, that source is noted. The word may be one that has dual meanings or one that could possibly be confusing or hard to understand for an English learner. The definitions generally come from the Student Glossary, found in the Home Connections booklet, and from the wording used in the lessons.

In the Guide a word or phrase for each lesson is selected and defined, and a brief activity is given to help the child develop an understanding of the meaning. While often reflecting an activity completed in the class, the emphasis should be on conveying the meaning of the vocabulary word, not the activity itself. Since many children in this group will be Spanish speakers, Spanish cognates (words of the same meaning and similar in sound) are listed. This will help the child connect the meaning of the words in the two languages.

Tip

Four-Step Approach:

1. Define the word.
2. Model the meaning.
3. Check for understanding.
4. Apply in context.

A four-step approach is reflected in the English Learner activities:

1. Define the word: the child *hears* the word and is given a description of its meaning.
2. Model the meaning of the word: the child *sees* and *connects* the word to its meaning.
3. Check for understanding: the child *talks about* and *explains* the term.
4. Apply in context: the child *uses* the word.

It is suggested that the words be written on the board and frequently repeated. A collection of index cards with the words and meanings or a copy of the Student Glossary may be helpful. Gestures, pictures, restating, and paraphrasing may aid in conveying meanings. To prevent the child from feeling isolated, pair the English learner with a friend who can help when possible.

Other resources for English learners include:

- *Saxon Math* English Learners Handbook, which includes translations in nine languages, Spanish cognates, and information regarding grammar and phonics transfer issues

- Student Glossary (in the Home Connections booklet located in the Monitoring Student Progress binder), which includes a Spanish translation of mathematical terms and definitions
- Teacher Glossary (in the Teacher's Manual, Volumes 1 and 2), which provides a Spanish translation of mathematical terms and definitions used throughout the year
- Extend and Challenge CD activities provided in Spanish
- Complete program available in Spanish.

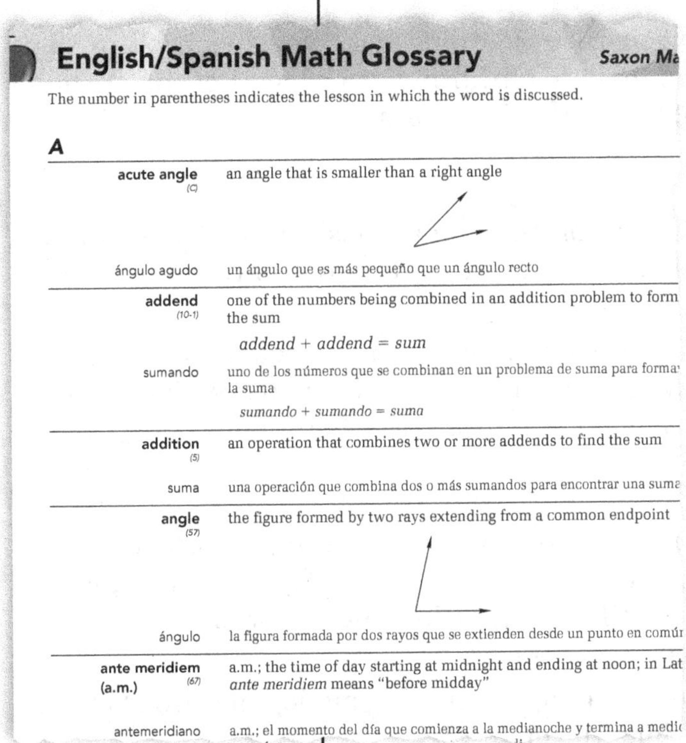

English/Spanish Math Glossary

Saxon Ma

The number in parentheses indicates the lesson in which the word is discussed.

A

acute angle (C) — an angle that is smaller than a right angle

ángulo agudo — un ángulo que es más pequeño que un ángulo recto

addend (10-1) — one of the numbers being combined in an addition problem to form the sum

addend + *addend* = *sum*

sumando — uno de los números que se combinan en un problema de suma para forma la suma

sumando + *sumando* = *suma*

addition (5) — an operation that combines two or more addends to find the sum

suma — una operación que combina dos o más sumandos para encontrar una suma

angle (57) — the figure formed by two rays extending from a common endpoint

ángulo — la figura formada por dos rayos que se extienden desde un punto en comúr

ante meridiem (a.m.) (67) — a.m.; the time of day starting at midnight and ending at noon; in Lat *ante meridiem* means "before midday"

antemeridiano — a.m.; el momento del día que comienza a la medianoche y termina a medi

Inclusion Students

Inclusion tips are intended for teachers of children who may have difficulty with distractibility, visual-motor integration, fine-motor coordination, short-term memory, number reversals, verbal explanation, spatial orientation, or receptive language. Children with special needs can make good progress in mathematics when modifications are made according to their strengths and needs and when the lesson content is accessible. Because the type of disability (physical, intellectual, or emotional) and degrees of limitation vary too widely to give specific instruction to meet each need, general and modified activities have been created. It may be necessary after evaluating the child to further adjust an activity to meet his or her specific need.

At the beginning of the school year, it is important to review the child's assessments from the previous year and to talk with other teachers about any modifications they made. If the child has an Individual Education Plan (IEP), review the plan carefully with the special education teacher.

Tip

Children with special needs can make good progress in mathematics when modifications are made according to their strengths and needs.

Following are some general guidelines and some specific tips that have helped teachers include children with special needs in the different sections of the *Saxon Math* lessons:

Classroom Seating

- When children are working at their desks or tables, seat the child with special needs close to the front of the class and/or close to you.
- When children are seated in a group or circle on the floor, seat the child with special needs close to you, which allows you to monitor behavior and/or participation.

The Meeting

- Modify questions in The Meeting by asking a simpler question. For example, if the question asks what the date was one week ago, modify the question and ask what the date was one day ago.
- Provide additional calendar, time, or money questions at the child's instructional level.
- Encourage the child to answer questions he or she knows.
- Encourage the child to participate in the group responses.

- When practicing skip counting, make sure the counting strips are visible to all children. Point to the numbers as the children count.

Fact Practice

- Introduce fewer facts for the child to learn at one time.
- Circle the problems the child should do first on his or her fact sheet.
- Have the child use a one-row viewer. (See Teachers Resource Handbook, page 12.)
- Use fact sheets from earlier lessons or from a previous grade, or make modified fact sheets. Modify to fit a child's needs.
- Reduce the number of problems by cutting or folding the sheet.
- Place a fact on the door and have the child say that fact each time he or she goes in or out.
- Have another child help the child with special needs practice the facts orally.
- Allow the child to take as much time as needed to complete a fact sheet and record the amount of time needed to complete it. On the following days have the child try to complete the fact sheet in less time.

New Concepts

- Have the child restate the directions or problem in his or her own words.
- To help a child who has trouble processing instructions, give one direction at a time. Pause and then give the next direction.
- Simplify an example or give additional examples.
- Have the child use manipulatives to solve a problem.

New Concepts *Gaining Knowledge: a Mathematical Conversation*

- **Adding 10 to a Multiple of 10**
- **Finding Missing Numbers on a Piece of the Hundred Number Chart**

"Today you will learn how to add 10 to one of these numbers."

- Point to the multiples of 10 on the hundred number chart or the cou by 10's counting strip.

"You will also learn how to find missing numbers on a piece of a hundred number chart."

"Take out your Math Folder."

Math Folder

"Put your Math Folder on your desk so you can see the hundrec number chart."

- Allow time for the children to do this.

"Point to 20 on the hundred number chart."

"Point to the number that is ten more."

"What number is ten more than 20?" 30

- Make a vocabulary word wall and review words as needed. (Use English Learner tips when applicable.)
- Have the child build his or her mathematics vocabulary by creating a personal "Math Dictionary" notebook.
- Give the child a copy of the Student Glossary found in the Home Connections in the Monitoring Student Progress binder.

Guided Class Practice

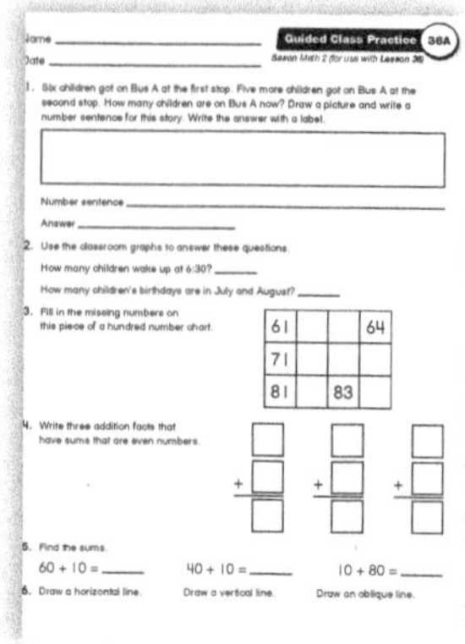
Name ______ Date ______

Guided Class Practice 36A

Saxon Math 2 (for use with Lesson 36)

1. Six children got on Bus A at the first stop. Five more children got on Bus A at the second stop. How many children are on Bus A now? Draw a picture and write a number sentence for this story. Write the answer with a label.

Number sentence ______

Answer ______

2. Use the classroom graphs to answer these questions.

How many children wake up at 6:30? ______

How many children's birthdays are in July and August? ______

3. Fill in the missing numbers on this piece of a hundred number chart.

61			64
71			
81		83	

4. Write three addition facts that have sums that are even numbers.

5. Find the sums.

60 + 10 = ______ 40 + 10 = ______ 10 + 80 = ______

6. Draw a horizontal line. Draw a vertical line. Draw an oblique line.

- If a child has difficulty reading, read aloud all directions and problems to him or her.
- Circle or underline key words or ideas on the child's Guided Class Practice.
- For a child who has difficulty following multi-step directions, mark directions with different color highlighters.
- Have the child restate the directions or problem in his or her own words.
- To help a child who has trouble processing instructions, give one direction at a time. Pause and then give the next direction.
- Number the directions in multi-step problems.
- If possible, simplify an example or give additional examples.
- Have the child use manipulatives to solve a problem.
- Allow the child to give answers orally and record the answers for the child.
- Provide additional time for the child to complete the Guided Class Practice.

Assessment

- Modify the problems on the assessment if necessary.
- If a child has difficulty reading, read aloud all directions and problems to him or her.
- Include previous modifications utilized that helped the child meet with success.
- Use the Reteaching Activities and the Reassessment Questions found on the *Saxon Math* eGradebook CD.

At-Risk Students

Children are considered at risk when they have the capability of successfully completing the coursework presented, but have fallen behind due to various circumstances (such as illness or other problems).

The activities for at-risk students contain review and reinforcement projects that are modifications of the material covered in the lesson. The New Concept of each lesson is the primary source of the mathematical concepts; it is therefore beneficial to go back to the lesson in which the concept was first presented, if there is ongoing difficulty in understanding.

Included in the activities for at-risk students are the Math Center Activities. A complete listing of these activities is in the Math Center Activities booklet in the Monitoring Student Progress binder. These activities are useful for the entire class, but are especially suitable for at-risk children. Some of the activities in this section are taken from the Math Center Activities booklet and are designed for four children at a time but can be used for an individual child or modified for numbers other than four.

Some activities in the Math Center involve practice with Learning Wrap-Ups®. Included with the *Saxon Math* program, Learning Wrap-Ups provide each student with a tactile means of practicing number facts. Speed and accuracy are promoted by frequent use of these mathematical tools.

This Guide lists one activity per lesson for an at-risk student, but some lessons in the Math Center Activities booklet have more than one activity. In those cases, one is usually a Learning Palette®[1] activity and the other is an activity reviewing a concept taught in the lesson. Learning Palettes are self-correcting learning centers that provide independent hands-on practice, reinforcement, and enrichment of math concepts. Learning Palettes are an optional purchase and come packaged for use by up to four children at a time. The content is written for the *Saxon Math* program and supports the concepts taught in the lessons.

The Extend and Challenge CD is an excellent tool to help children at risk practice and review concepts. This CD includes both English and Spanish versions of the interactive activities for Kindergarten through Grade 3 and can be used independently by

[1] Learning Palettes are available for *Saxon Math 1–3* and are an optional purchase. They may be ordered by calling Customer Service at 800.531.5015

most children. Although the CD activities are written to extend and challenge children, many of the activities have differing difficulty levels and are suitable for at-risk children. Carefully reviewing the possibilities the CD offers will help the teacher assign the right activity at the right time to meet a child's need.

Other resources for at-risk students include:

- Reteaching Activities and Reassessment Questions found on the *Saxon Math* eGradebook CD
- Online activities for each grade, both general and lesson specific, at the Student Activity Center at www.SaxonMathActivities.com.

Advanced Learners

Advanced learners create a pleasant challenge for the teacher. These children quickly and easily understand the concepts of the lessons, but may require additional stimulation to continue their enthusiasm for learning.

Some of the activities for advanced learners are the Extend and Challenge CD activities. This CD includes both English and Spanish versions of all of the interactive activities for Kindergarten through Grade 3 and can be used independently by most children. The activities have difficulty levels and can be used by students at grade level as well as remediation activities for grade levels above or advanced learning for grade levels below. The multiple uses of this CD can vary greatly, and offer children at all levels of learning the use of technology. (Note: The voice on the CD tells the children to click the Done button on the screen; on the shopping activities click the Done button or use the Enter key on the keyboard, if necessary.)

There are two versions of the Extend and Challenge booklet; the one included in the Monitoring Student Progress binder provides recording forms and worksheets in English, while the stand-alone version provides these in both English and Spanish. The Extend and Challenge CD and reference guide are located inside the cover of the stand-alone version.

Ideas for Journal Writing (which the child may complete independently) are also presented in the Advanced Learners section. Additional writing suggestions include:

- Asking the child how new skills or concepts may be used in real life.
- Asking the child to make up a story or problem similar to those used in the lesson.
- Asking open-ended questions, such as, "What do you think would happen if ...?"

This Guide provides some activities for extending learning through additional problem solving. Some of the problems are taken from the Extension and Enrichment problems available at www.SaxonMathResources.com. These problems were adapted and extended from Guided Class Practice Problems and from Problem-Solving Worksheets. (Note: If you choose to print the problem from the Web site, select Current page for printing or adjust the page number to account for the introductory pages.)

Learning Palette Enrichment cards also provide challenging Advanced Learners activities. The Learning Palettes are an optional purchase and come packaged for use by up to four children at a time. The Enrichment cards are provided in a package included in the set and are intended for individual use. The palettes provide self-checking so that each activity can be done independently by the child.

For most lessons only one activity has been listed for Advanced Learners. Other suggestions may be found on the last page of the Overview for each set of lessons.

Additional resources for advanced learners include:

- Online activities for each grade, both general and lesson specific, at the Student Activity Center at www.SaxonMathActivities.com.

Guide to Differentiated Instruction

- **Reading and Identifying Numbers to 100**
- **Identifying Right and Left**

English Learners

materials: Math Folder (Left and Right hands)

- Show the child the left and right hands on the Math Folder. (The child may use this for reference.)
- **left:** a direction; the opposite of right
- **right:** a direction; the opposite of left
- Remind him or her how an *L* is formed with the left hand (as shown) and that this is a good hint to remember which hand is which.
- Point to the child's right hand and ask him or her which hand it is.
- Repeat for the left hand.

Inclusion Students

The child will identify right and left.

materials: Math Folder (Left and Right hands)

- Show the child the left and right hands on the Math Folder. (The child may use this for reference.)
- Remind him or her how an *L* is formed with the left hand (as shown) and that this is a good hint to remember which hand is which.
- Ask the child to: touch his or her right ear, raise his or her left hand, point to his or her left foot, put a hand on his or her right knee.
- Point to the child's right hand and ask him or her which hand it is.
- Repeat for the left hand.

At-Risk Students

- Math Center Activity 1

Objectives	Materials for Four Children	Teacher Directions
• Identifying Numbers to 100	• 4 Learning Palettes • 4 Learning Palette Card 1	Ask each child to match the problems and answers on Learning Palette Card 1.

Advanced Learners

The child will read and identify numbers to 100.

materials: Math Folder (hundred number chart) and pennies (or other small tokens)

- Place pennies over the following numbers on the child's hundred number chart: 4, 16, 29, 32, 40, 54, 57, 66, 81, 99.
- Ask the child to identify which numbers are hidden; as each number is named remove the coin to confirm the answer.

- **Identifying One More and One Less Than a Number**
- **Graphing Data on a Graph**

English Learners

materials: Math Folder (hundred number chart)

- Show the child the numbers on the hundred number chart and review that they are all the numbers in order from 1 to 100.
- Explain that every number on the chart is made from one or more of the digits 0 to 9.
- **number:** one or more digits showing an amount
- Speaking slowly, if necessary, ask the child to point to the following numbers: 33, 58, 6, 25, 13, 82, 60, 47.
- Point to the following numbers and ask the child to say them: 95, 13, 56, 7, 48, 82, 21, 12.
- Ask the child to read the numbers in the 71–80 row on the hundred number chart.
- Point to the hundred number chart and ask the child what the characters are called.

Spanish Cognate: number número

Inclusion Students

The child will identify one more and one less than a number, using left and right directions.

materials: Math Folder (hundred number chart)

- Have the child point to and identify the following numbers: the number to the right of 61, to the right of 45, to the left of 50, to the left of 18, to the right of 55, to the left of 32, to the right of 32.

At-Risk Students

- Math Center Activity 2

Objectives	Materials for Four Children	Teacher Directions
• Identifying Numbers That Are One More, One Less, and Between	• 4 Learning Palettes • 4 Learning Palette Card 2	Ask each child to match the problems and answers on Learning Palette Card 2.

Advanced Learners

materials: Math Folder (hundred number chart) and pennies (or other small tokens)

- Place pennies over the following numbers on the child's hundred number chart: 3, 18, 22, 35, 45, 53, 60, 68, 69, 71.
- Ask the child to say the number under the first coin.
- Ask him or her to identify what number is one more than and one less than the covered number.
- Repeat with the remaining numbers.

- **Telling and Showing Time to the Hour**

English Learners

materials: demonstration clock

- Point to the two hands on the demonstration clock; review that the long hand moves around the clockface once each hour, and in that period the short hand will move from one number to the next, showing the hour.
- **hour hand:** the shorter hand on the clock that shows the hour
- **minute hand:** the longer hand on a clock that shows the number of minutes before and after the hour
- Set the demonstration clock to 4:00.
- Ask the child to say the time.
- Repeat with the following times: 9:00, 12:00, 2:00.
- Point to the hour hand on the clock and ask the child to identify it.
- Repeat with the minute hand.

Inclusion Students

materials: student clock

- Set the student clock to 6:00; ask the child to say the time.
- Repeat with the following times: 2:00, 12:00, 5:00.
- Write 11:00 on the board.
- Ask the child tell to you what time it is and to set the clock to match.
- Repeat with the following times: 8:00, 4:00, 3:00.

At-Risk Students

materials: student clock

- Set the student clock to 6:00; ask the child to say the time.
- Repeat with the following times: 2:00, 12:00, 5:00.
- Ask the child to set the clock to the following times: 10:00, 1:00, 5:00, 7:00.
- Write 11:00 on the board; ask the child to tell you what time it is and to set the clock to match.
- Repeat with the following times: 8:00, 4:00, 3:00.

Advanced Learners

materials: student clock

- Ask the child to set the clock to 10:00.
- Then ask him or her to set the clock one hour later.
- Repeat with 1:00.
- Ask the child to set the clock to 5:00.
- Then ask him or her to set the clock one hour earlier.
- Repeat with 7:00.

- **Writing Numbers to 100**

English Learners

materials: Math Folder (hundred number chart)

- Using the Math Folder, point to 13 to 19 as you read the numbers with the child.
- **-teen:** ten, when it is part of a counting number
- Using the hundred number chart as reference, ask the child to give the answer as you say the following: 3 + 10 = , 4 + 10 = , 5 + 10 = , and so on to 9 + 10 =.
- Remind the child that all the teen numbers begin with a *1*.
- Ask the child to say the numbers from 13 to 19. Point out that 13 and 15 are spoken a little differently than just adding *teen* to the number.

Inclusion Students

- Ask the child to listen carefully as you say the following numbers and have him or her write each one as you say it: 65, 41, 77, 83, 26, 19, 30, 96.
- Enunciate and emphasize the parts of the number when possible, such as *six*ty-*five*.

At-Risk Students

- Math Center Activity 3

Objectives	Materials for Four Children	Teacher Directions
• Counting by 10s	• 4 Learning Palettes • 4 Learning Palette Card 3	Ask each child to match the problems and answers on Learning Palette Card 3.

Advanced Learners

- Extend and Challenge CD, Activity 1, Find the Missing Number
- After the child finishes the activity, use the question(s) on the recording form to evaluate his or her understanding. (Monitoring Student Progress binder, Extend and Challenge CD booklet)

- **Addition Facts: Doubles with Sums to 18**

English Learners

materials: Teacher Fact Cards (Doubles)

- Show several doubles fact cards.
- **doubles:** two of the same number
- Explain that when describing addition facts, *doubles* means that the two numbers being added together are the same, such as 1 + 1, 2 + 2, 3 + 3, and so on.
- Ask the child to notice that on each card the two numbers being added together are the same.
- Have the child say the answer to 2 or 3 of the doubles addition facts.
- Ask the child what the addition fact problems are called when the numbers being added are the same.

Spanish Cognate: doubles dobles

Inclusion Students

materials: Teacher Fact Cards (Doubles, 1 + 1 through 4 + 4)

- Show the Teacher Fact Cards in random order; ask the child to say the answers.
- Review any facts the child misses.

At-Risk Students

materials: Teacher Fact Cards (Doubles, 1 + 1 through 9 + 9)

- Show the Teacher Fact Cards in random order; ask the child to say the answers.
- Review any facts the child misses.

Advanced Learners

- Online Activity, Doubles
  (Student Activity Center is available at www.SaxonMathActivities.com.)

• Identifying the Attributes of Pattern Blocks

English Learners

materials: Math Folder (Pattern Block Polygons) and pattern blocks (1 of each color)

- Show the child the pattern blocks and the pattern block descriptions on the Math Folder.
- **shape:** a closed form or outline
- Explain to the child that these are the shapes that were used in class.
- Using the Math Folder, ask the child to place each pattern block over its outline.
- Ask him or her to tell you how the 6 shapes are different. (size, color, number of sides, etc.)
- Ask the child to use a complete sentence to explain *shape*.

Inclusion Students

The child will identify the attributes of three pattern blocks.

materials: 18 pattern blocks (6 each of the blue, green, and orange)

- Give the child the blue, green, and orange pattern blocks.
- Ask the child the colors of the three different pattern blocks and the number of sides on each.
- Ask him or her to sort the blocks and place each color in a different row.
- Review the names of the three types of blocks. (parallelogram, triangle, square)

At-Risk Students

- Math Center Activity 4

Objectives	Materials for Four Children	Teacher Directions
• Free Exploration of Pattern Blocks	• basket of pattern blocks	Allow children to explore the pattern blocks.

Advanced Learners

Journal Writing (Overview 1)

materials: pattern blocks

- Have the child choose 1 pattern block and trace it; ask him or her to create a picture using the pattern block and to write about the picture.

- **Identifying Ordinal Position to Sixth**
- **Creating and Reading a Repeating Pattern**

English Learners

materials: pattern blocks (1 of each color)

- Show the child the blocks arranged in the following order: green, red, yellow, blue, orange, and tan.
- **ordinal number:** a number that describes position or order
- Explain that the green pattern block is the first of the blocks.
- Point to each pattern block as you say the ordinal position—first, second, third, ... to sixth.
- Ask the child to point to the appropriate block as you ask the following: which block is second, ... fifth, ... first, ... last, ... fourth, ... sixth, ... third.
- Ask the child to use a complete sentence to explain *ordinal number.*

Spanish Cognate: ordinal number número ordinal

Inclusion Students

The child will identify ordinal position to the third.

materials: 3 index cards (labeled *first*, *second*, and *third*) and 3 color tiles

- Have the child place the color tiles in a row.
- Ask the child to place the index card labeled *third* under the *third* color tile.
- Repeat with the *first* and *second* cards.

At-Risk Students

- Math Center Activity 6

Objectives	Materials for Four Children	Teacher Directions
• Identifying the Next Shape in a Repeating Pattern	• 4 Learning Palettes • 4 Learning Palette Card 4	Ask each child to match the problems and answers on Learning Palette Card 4.

Advanced Learners

- Learning Palette Enrichment Card E1, Identifying the Missing Shape in a Repeating Pattern

- **Identifying and Acting Out "Some, Some More" Stories**
- **Comparing Numbers to 50**

English Learners

materials: 13 color tiles

- Show the child two rows of color tiles, one with 4 and one with 9.
- **greater:** more
- Ask him or her to point to the row with the greater number of tiles.
- Then ask the child how many tiles are in the longer row.
- Repeat with rows of 8 and 3.
- Ask the child to use a complete sentence to explain *greater*.

Inclusion Students

The child will compare numbers to 30.

materials: Math Folder (hundred number chart) and 2 color tiles

- Cover the 5 and 9 with the color tiles.
- Ask the child to identify which two numbers are covered and which number is greater.
- Repeat with the following pairs of numbers: 4 and 27; 5 and 25; 12 and 15; 10 and 28.

At-Risk Students

The child will compare numbers to 50.

materials: Math Folder (hundred number chart) and 2 color tiles

- Cover the 5 and 9 with the color tiles.
- Ask the child to identify which two numbers are covered and which number is greater.
- Repeat with the following pairs of numbers: 4 and 27; 5 and 25; 12 and 15.
- Write 17 and 50 on the board; ask the child to identify which number is greater, using the hundred number chart for reference, if necessary.
- Repeat with the following pairs of numbers: 18 and 34; 48 and 9; 26 and 13; 41 and 47; 20 and 21; 31 and 29.

Advanced Learners

The child will create and act out a some, some more story.

materials: color tiles

- Ask the child to make up a some, some more story.
- Have the child act out his or her some, some more story using the color tiles.

- **Comparing and Ordering Objects by Size (Area)**

English Learners

materials: pattern blocks (hexagon, trapezoid, parallelogram, and triangle)

- Show the child 1 of each of the pattern blocks and explain that they are all different sizes.
- **size:** a measurement of how large or how small something is
- Ask the child to arrange the four shapes in order from the smallest to the largest size.
- Ask the child to use a complete sentence to explain *size*.

Inclusion Students

materials: pattern blocks (hexagon, trapezoid, square, and triangle)

- Ask the child to arrange the four shapes in order from the largest to the smallest.
- If the child has trouble, ask the child to select the largest out of the four first, then the largest from the remaining three, and finally the larger of the two remaining shapes.

At-Risk Students

materials: pattern blocks (hexagon, trapezoids, and parallelograms)

- Ask the child to cover the yellow pattern block with red pattern blocks.
- Ask him or her how many red blocks are needed to cover the yellow pattern block. (2)
- Then ask the child to cover the yellow pattern block with blue pattern blocks.
- Ask him or her how many blue pattern blocks are needed to cover the yellow pattern block. (3)
- Ask the child to compare the number of blue and red pattern blocks needed to cover the yellow pattern block.
- Ask why more blue pattern blocks are needed than red pattern blocks; if needed, explain that the larger the pattern block, the fewer it takes to cover the bottom shape.
- Ask the child to place the 3 types of blocks in order from largest to smallest.

Advanced Learners

materials: pattern blocks

- Ask the child to cover the blue parallelogram with green triangles. (Keep all the shapes covered until the end of the activity.)
- Ask him or her to tell you how many he or she used. (2)
- Now ask the child to cover the red trapezoid with green triangles.
- Ask him or her to tell you how many he or she used. (3)
- Now ask the child to cover the yellow hexagon with green triangles.
- Ask him or her to tell you how many were used. (6)
- Ask the child to pretend that a green triangle is worth one penny.
- Ask him or her to tell you the value of the blue parallelogram, the red trapezoid, and the yellow hexagon. (2¢, 3¢, 6¢)

Guide to Differentiated Instruction

- **Addition Facts: Adding 0 and Adding 1**
- **Identifying Addends, Sums, and the Commutative Property of Addition**
- **Using Logical Reasoning to Solve a Problem**
- **Solving a Problem by Acting It Out**

English Learners

materials: 2 index cards; on one write the number sentence 4 + 1 = 5 and on the other write the same numbers as a vertical problem

- On the board write *addend* + *addend* = *sum.*
- Show the child the two cards.
- **addend:** one of the numbers being combined in an addition problem to form the sum
- **sum:** the answer to an addition problem
- Point out the addends and the sum on each card. Show that the direction of the problem does not change the name of the terms.
- Ask the child what the answer to an addition problem is called.
- Ask the child what the numbers being added are called.

Spanish Cognate: sum suma

Inclusion Students

materials: one-row viewer (instructions found in the Teacher Reference Handbook, pg. 12)

- For the child who lacks focus, have him or her redo Lesson Worksheet 10-1A/B using a one-row viewer

At-Risk Students

materials: Problem-Solving Worksheet 10A and Performance Task Worksheet 10B

- Review with the child the problem-solving strategies and Problem-Solving Worksheet 10A and Performance Task Worksheet 10B.

Advanced Learners

- Extension and Enrichment problem for Problem Solving Worksheet 10A (Extension and Enrichment Student Problem [page 1] at www.SaxonMathResources.com)

Extension and Enrichment Problem	Answer
Kayne placed his 5 shirts in his drawer. He placed the green shirt in the drawer first. Then he stacked the yellow, pink, black, and white shirts in the drawer. The yellow shirt is not on top. The pink shirt is above the white shirt. The black shirt is under the white shirt. The yellow shirt is on top of the green shirt. Draw a picture to show the stack and which shirt Kayne placed on top.	a drawing showing the stack of 5 shirts in this order: pink (on top), white, black, yellow, and green

- **Covering a Design Using Pattern Blocks**

English Learners

materials: pattern blocks

- Show the child the blocks arranged in the following order: green, red, yellow, blue, orange, and tan.
- **first:** the ordinal number that tells which one is before all the others
- **second:** the ordinal number that tells the number two position
- Explain that the green pattern block is the first of the blocks.
- Point to each pattern block as you say the ordinal position—first, second, third, ... to sixth.
- Ask the child to point to the appropriate block as you ask the following: which block is second, ... fifth, ... first, ... last, ... fourth, ... sixth, ... third.
- Point to each block and ask the child to name the ordinal position of the block.

Spanish Cognate: second segundo

Inclusion Students

- To help a child who has trouble processing instructions, give one direction at a time. Pause and then give the next direction.

At-Risk Students

- Math Center Activity 7

Objectives	Materials for Four Children	Teacher Directions
• Covering a Design Using a Given Number of Pattern Blocks	• children's completed Lesson Worksheet 10-2 (from Lesson 10-2)	Ask each child to select another child's Lesson Worksheet 10-2 and use the number of pattern blocks written on the paper to cover each design.

Advanced Learners

- Extend and Challenge CD, Activity 2, Cover the Shape
- After the child finishes the activity, use the question(s) on the recording form to evaluate his or her understanding. (Monitoring Student Progress binder, Extend and Challenge CD booklet)

• Identifying and Acting Out "Some, Some Went Away" Stories

English Learners

materials: pattern blocks (hexagon, trapezoid, square, and triangle)

- Point to the hexagon and the square, and show the child that the square is smaller than the hexagon.
- **smaller:** when comparing two objects, the lesser of the two
- Compare two pattern blocks and ask the child to select and state which is smaller.
- Repeat two more times using different pattern blocks.
- Have the child use a complete sentence to explain *smaller.*

Inclusion Students

materials: 8 color tiles

- Ask the child to use the color tiles to show what is happening in the following problem: Paige had 8 crayons. She gave 2 to her brother.
- Ask the child to identify the problem as a some, some more or a some, some went away story.
- Ask him or her how many crayons Paige had after she gave away 2.
- Repeat this activity with the following problem: Shawn had 7 toy cars; he gave 3 to his friend Jake.

At-Risk Students

materials: 8 color tiles

- Ask the child to use the color tiles to show what is happening in the following problem: Paige had 8 crayons. She gave 2 crayons to her brother.
- Ask the child to identify the problem as a some, some more or a some, some went away story.
- Ask him or her how many crayons Paige had after she gave away 2.
- Repeat this activity with the following problems: Shawn had 7 toy cars; he gave 3 to his friend Jake. Lori had 6 cookies; she gave 2 to her brother; then she gave 2 to her sister.
- Ask the child to make up his or her own some, some went away story, using the color tiles to act it out.

Advanced Learners

materials: color tiles

- Ask the child to explain the difference between a some, some more story and a some, some went away story.
- Ask him or her to use the color tiles to act out one of each story.

- **Identifying the Time One Hour Ago and One Hour From Now**
- **Numbering a Clockface**

English Learners

materials: student clock

- Write 1:00 on the board; remind the child that this is a digital time.
- **digital time:** the time shown on a clock using digits instead of using a dial and hands
- Ask the child to say and set the clock to this time; repeat with 5:00 and 9:00.
- Set the clock to 2:00; ask the child to say the time and write the digital time; repeat with 4:00 and 10:00.
- Ask the child to use a complete sentence to explain the term *digital time.*

Inclusion Students

The child will identify the time one hour ago and one hour from now.

materials: 2 student clocks

- On one clock, show 3:00. Tell the child that a baker has put a pie into the oven and it will be done in one hour. Ask the child to show (on the 2nd clock) what time the pie will be done.
- On one clock, show 7:00. Tell the child that Aaron will practice the piano for one hour. Ask the child to show (on the 2nd clock) what time Aaron will be done practicing.
- On one clock, show 1:00. Tell the child that Molly began cleaning the garage one hour ago. Ask the child to show (on the 2nd clock) what time Molly began cleaning the garage.

At-Risk Students

The child will identify the time one hour ago and one hour from now.

materials: student clock

- On the board write 10:00.
- Have the child write the digital time on the board that is one hour after the time shown and then set the clock to the new time.
- Repeat with: 1:00, 11:00, 4:00.
- On the board write 2:00 and ask the child to set the time on the clock one hour before the time shown.
- Repeat with: 6:00, 9:00, 4:00.

Advanced Learners

- Extension and Enrichment problem for Guided Class Practice 12, problem 1 (Extension and Enrichment Student Problem [page 1] at www.SaxonMathResources.com)

Extension and Enrichment Problem	Answer
There were 8 children on the bus. At the next bus stop, 3 more children got on the bus. At the next stop, 1 more child got on the bus. How many children are on the bus now?	12 children

• Identifying Even and Odd Numbers

English Learners

materials: Math Folder (hundred number chart) and 10 pennies

- **even:** every number divisible by 2; every number ending in 0, 2, 4, 6, or 8
- **odd:** every number not divisible by 2; every number ending in 1, 3, 5, 7, or 9
- Ask the child to put a penny on the first 10 odd numbers on the hundred number chart.
- When he or she is done, repeat with the first 10 even numbers.
- Have the child use a complete sentence to explain *even* and *odd*.

Inclusion Students

materials: index cards numbered from 0 to 15

- Have the child sort the numbers into even or odd.

At-Risk Students

- Math Center Activity 8

Objectives	Materials for Four Children	Teacher Directions
• Identifying the Missing Number in a Sequence of Even or Odd Numbers	• 4 Learning Palettes • 4 Learning Palette Card 5	Ask each child to match the problems and answers on Learning Palette Card 5.

Advanced Learners

- Learning Palette Enrichment Card E2, Identifying the Missing Number in a Sequence of Even and Odd Numbers

- **Identifying Ordinal Position to Twelfth**

English Learners

materials: the Meeting Board Today's Pattern for Lesson 14 and color tiles (6 blue and 6 yellow)

- Make an AB color pattern beginning with yellow, and show the child the 12-tile color pattern.
- **tenth:** the ordinal number that tells the number ten position
- Ask him or her what color the 10th tile is. (blue)
- Refer to Today's Pattern.
- Ask the child which figure is in the 10th position. (hexagon)
- Have the child use a complete sentence to explain *tenth*.

Inclusion Students

materials: 12 index cards; write *first* on one side of an index card and *1st* on the other side; repeat with *second* through *twelfth*

- Give the child the index cards number side up.
- Ask him or her to place the cards in order from 1st to 12th.
- Ask the child to repeat the activity with the word side up; assist as needed.
- Have the child check the final order by turning the cards back over to the number side.

At-Risk Students

- Math Center Activity 10

Objectives	Materials for Four Children	Teacher Directions
• Identifying Ordinal Position to Twelfth	• 4 Learning Palettes • 4 Learning Palette Card 6	Ask each child to match the problems and answers on Learning Palette Card 6.

Advanced Learners

- Ask the child to give examples of ordinal numbers in everyday life. (Examples: street names, floors in buildings, bases in baseball, prizes in contests, ranking of sport teams, etc.)

- **Addition Facts: Adding 2**

English Learners

- **commutative property of addition:** the rule that states that changing the order of the addends does not affect the sum
- Review **addend:** one of the numbers being combined in an addition problem to form the sum
- Review **sum:** the answer to an addition problem
- Write 2 + 4 = _____ on the board and ask the child for the sum.
- Directly under this, write 4 + 2 = _____ and ask the child for the sum.
- Then draw a two-headed arrow between the 4's; draw a second arrow between the 2's.
- Explain that commutative property of addition shows that since the addends themselves did not change, only their order, the answer will be the same.
- Write 8 + 2 = _____ + _____ on the board and have the child demonstrate the commutative property of addition by switching the addends to fill in the blanks; ask what the sum is.
- Have the child use a complete sentence to explain *commutative property of addition.*

Spanish Cognate: sum suma

Inclusion Students

materials: Teacher Fact Cards (Set A, Adding 2: use only the cards on which 2 is the upper number) and color tiles (two colors)

- Show the Teacher Fact Cards in random order; ask the child to say the answers.
  Note: If the child has difficulty, allow him or her to use the tiles to act out the problem. Have the child count out the tiles (one color) for the first number in the fact; then count out tiles (second color) for the second number in the fact. Direct the child to count the total number of tiles to solve the problem.

Advanced Learners

materials: Teacher Fact Cards (Set A, Adding 2: remove the 2 + 2 and 0 + 2 cards) and color tiles

- Ask the child to pair any cards that show matching addition facts, that is, cards that show the commutative property of addition, such as 4 + 2 and 2 + 4.
- Ask the child to give you the sum of each card. If the child has difficulty, allow him or her to use color tiles to act out the problem.

- **Creating and Reading a Repeating Pattern**

English Learners

materials: pattern blocks

- Remind the child what a repeating pattern is by creating an AB pattern using pattern blocks and reading it together.
- **repeating pattern:** an arrangement of items that is shown over and over
- Create an ABC pattern and have the child continue it and then read the pattern.
- Have the child use a complete sentence to explain *repeating pattern.*

Spanish Cognate: repeat repetición

Inclusion Students

materials: pattern blocks (squares and triangles)

- Using the pattern blocks, create an AB pattern and have the child continue it and then read the pattern.
- Repeat this activity using an AAB pattern.

At-Risk Students

- Math Center Activity 12

Objectives	Materials for Four Children	Teacher Directions
• Identifying the Missing Shapes in a Repeating Pattern	• 4 Learning Palettes • 4 Learning Palette Card 7	Ask each child to match the problems and answers on Learning Palette Card 7.

Advanced Learners

materials: color tiles

- Using the color tiles, create an ABC pattern; ask the child to identify the pattern.
- Repeat this activity using the following patterns: AABB, AABC, ABBCDD.
- Have the child create a repeating pattern using at least 3 colors; ask him or her to describe the pattern.

• Identifying Weekdays and Days of the Weekend

English Learners

materials: Meeting Board calendar

- Point to the calendar and identify the weekdays and the weekend days; note that on both the Meeting Board calendar and Days of the Week list, weekend days are shown in yellow.
- **weekday:** Monday, Tuesday, Wednesday, Thursday, or Friday
- **weekend:** Saturday or Sunday
- Have the child identify the days he or she comes to school.
- Have the child identify the days he or she doesn't go to school.
- Point to Sunday, Friday, Monday, Saturday, and Thursday on the calendar and have the child identify them as weekdays or weekend days.
- Have the child use a complete sentence to explain *weekday* and *weekend.*

Inclusion Students

materials: Meeting Board calendar

- Have the child point to and name the days he or she comes to school.
- Review that these five days are called weekdays.
- Have the child point to and name the days he or she doesn't come to school.
- Review that these two days are called weekend days, and that they appear on either end of a calendar week.

At-Risk Students

- Math Center Activity 13

Objectives	Materials for Four Children	Teacher Directions
• Identifying a Day of the Week on a Calendar	• 4 Learning Palettes • 4 Learning Palette Card 8	Ask each child to match the problems and answers on Learning Palette Card 8.

Advanced Learners

materials: page from a calendar

- Have the child count the number of weekdays in the month.
- Then have the child count the number of weekend days in the same month.
- Ask him or her to add the amounts to find the total number of days in the month.

Guide to Differentiated Instruction Lesson 17

- **Creating and Reading a Pictograph**
- **Drawing a Pictograph**

English Learners

materials: Our Favorite Day of the Week graph

- Point to the graph and remind the child that this is a pictograph.
- **pictograph:** a way of showing information using pictures or symbols
- Explain that graphs are a useful way to show information; they don't need many words because they use pictures, lines, and so on.
- Ask the child what the tags on this graph show.
- Ask the child to tell you what type of graph this is. (pictograph)

Spanish Cognates: pictograph pictograma

Inclusion Students

The child will read a pictograph.

materials: Our Favorite Day of the Week graph

- Ask the child which is the favorite day of the week for the class.
- Ask the child which is the least favorite day of the week for the class.

At-Risk Students

The child will read a pictograph.

materials: Our Favorite Days of the Week graph

- Ask the child which is the favorite day of the week for the class.
- Ask the child which is the least favorite day of the week for the class.
- Ask him or her additional questions related to the graph. (Examples: How many children chose the first day of the week as their favorite day? How many more children chose Sunday than chose Tuesday? How many chose Friday? Are there any days that an equal number of children chose?)

Advanced Learners

- Extension and Enrichment problem for Guided Class Practice 17, problem 3 (Extension and Enrichment Student Problem [page 2] at www.SaxonMathResources.com)

Extension and Enrichment Problem	Answer
A shape pattern is triangle, square, triangle, circle, triangle, square, triangle, circle, and so on. If the shape pattern keeps repeating, what will the 30th shape be? The triangles are in the 1, 3, 5, and 7 positions. What type of numbers are these? The squares and circles are in the 2, 4, 6, and 8 positions. What type of numbers are these?	square; odd; even

• Identifying Polygons

English Learners

- Draw a rectangle and an obtuse triangle on the board.
- Explain that these are examples of polygons and that polygons only have straight lines.
- **polygon:** a closed, flat figure bounded by line segments
- Draw several squares, parallelograms, triangles, and various nonpolygon figures (such as, circles) in random order on the board.
- Have the child point to all the polygons.
- Have the child use a complete sentence to explain *polygon.*

Spanish Cognate: polygon polígono

Inclusion Students

materials: draw several squares, parallelograms, triangles, and various nonpolygon figures (such as, circles) in random order on a sheet of paper

- Review that polygons are closed figures that only have straight lines.
- Ask the child to color the polygons red and to trace the figures that are not polygons blue.

At-Risk Students

- Online Activity, Polygons and Nonpolygons
  (Student Activity Center is available at www.SaxonMathActivities.com.)

Advanced Learners

- Draw a regular octagon and tell the child that this specific polygon is called an octagon.
- Ask the child to count the number of sides. (8)
- Ask the child if he or she knows what common sign has this shape. (If the child does not immediately know the answer, give the following clues: there is a word with four letters on the sign, the sign is red, it is a traffic sign.) (stop sign)

• Identifying Fractional Parts of a Whole

English Learners

materials: Math Folder (Fractions), 1 blue parallelogram, and 3 green triangles

- Draw a circle on the board, divide it in half, and shade one half; point out the fractions on the right pocket of the Math Folder.
- **half:** one of two equal parts
- Explain that when a whole object, like a circle, is cut into two equal parts, each part is a half.
- Ask the child to cover the blue parallelogram with green triangles.
- Ask how many green triangles it took to cover the parallelogram. (2)
- Explain that each triangle is half of the parallelogram.
- Have the child use a complete sentence to explain *half.*

Inclusion Students

materials: 6 color tiles (orange)

- Create a 3 × 2 rectangle with the six orange color tiles.
- Ask the child to tell you how many squares make up one-half of the rectangle, how many make up one third of the rectangle, and how many make up one sixth of the rectangle.

At-Risk Students

materials: pattern blocks

- Have the child cover the hexagon with trapezoids.
- Ask the child how many trapezoids were used to cover the hexagon and what fractional part of the hexagon each trapezoid represents. (2, a half)
- Have the child cover the hexagon with parallelograms.
- Ask the child how many parallelograms were used to cover the hexagon and what fractional part of the hexagon each parallelogram represents. (3, a third)
- Have the child cover the hexagon with triangles.
- Ask the child how many triangles were used to cover the hexagon and what fractional part of the hexagon each triangle represents. (6, a sixth)

Advanced Learners

Journal Writing (Overview 2)

- Ask the child to draw a candy bar and then show how to divide it in half.
- Have the child write about who he or she would share it with.

- **Adding 10 to a Single-Digit Number**
- **Addition Facts: Adding 9**
- **Using Logical Reasoning to Solve a Problem**
- **Solving a Problem by Acting It Out**

English Learners

materials: classroom number line and Math Folder (hundred number chart)

- Point to the number line and tell the child that the numbers 0 through 9 are only one digit.
- **single-digit number:** a number that has only one digit
- Remind the child that in class you discussed that single-digit numbers have a number only in the ones' place. To add 10 to this number, you put a 1 in the tens' place.
- Show examples of this on the hundred number chart.
- Have the child use a complete sentence to explain *single-digit number.*

Inclusion Students

The child will solve a problem by acting it out.

materials: color tiles (yellow, red, green, and blue) and on a piece of paper write 1st day, 2nd day, 3rd day, 4th day in one line.

- Ask the child to put a pattern block, which represents the color of each sweater, by the day that Ella wears it.
- Read each sentence slowly as the child acts out the story:
  Ella visited her grandmother for four days.
  Each day she wore a different color sweater.
  On the first day she wore blue.
  On the last day she wore red.
  On the day after she wore blue, she wore green.
  On what day did she wear yellow?

At-Risk Students

materials: Problem-Solving Worksheet 20A and Performance Task Worksheet 20B

- Review with the child the problem-solving strategies and Problem-Solving Worksheet 20A and Performance Task Worksheet 20B.

Advanced Learners

Journal Writing (Overview 2)

- Ask the child to write a paragraph explaining the adding 9 trick.

- **Creating a Color Pattern**

English Learners

materials: the Meeting Board Today's Pattern for Lesson 20-2 and a pattern block (parallelogram)

- **parallelogram:** a shape with four straight sides that has two pairs of parallel lines (see Lesson 20-2, Today's Pattern)
- Explain that the sides next to each other do not have to equal each other in length; show that on the pattern block they are equal, but in the drawing on Today's Pattern they are not; emphasize that it is the shape that matters.
- Draw the following on the board: trapezoid, triangle, parallelogram, circle, hexagon.
- Ask the child to identify the parallelogram.
- Have the child use a complete sentence to describe a *parallelogram.*

Spanish Cognate: parallelogram paralelogramo

Inclusion Students

The child will create a color pattern.

materials: color tiles (two or three colors) and on a sheet of paper, draw a 6" × 4" grid

- Have the child make an ABB pattern on the 6 × 4 grid using the color tiles.
- Remove the tiles and rotate the grid so that it becomes a 4 × 6 grid.
- Have the child again make an ABB pattern.
- Repeat the two steps with an ABC pattern, if desired.

At-Risk Students

The child will create a color pattern.

materials: a grid on a sheet of paper that is 7 columns across and five rows down, and crayons

- Have the child write each letter of his or her name in a box.
- Have the child continue writing his or her name until all the boxes on the grid are filled.
- Have the child use a crayon to color each letter a different color.
- Ask the child what he or she notices about the pattern.

Advanced Learners

- Extend and Challenge CD, Activity 3, Looking for Patterns
- After the child finishes the activity, use the question(s) on the recording form to evaluate his or her understanding. (Monitoring Student Progress binder, Extend and Challenge CD booklet)

• Identifying and Sorting Common Geometric Shapes by Attribute

English Learners

materials: Attribute Shape pieces from Lesson 21 (two colors)

- Sort all the pieces by size and show the child.
- **sort:** to separate items into groups based on something they have in common
- Explain that objects may be sorted by size, color, shape, or for any other reason decided upon.
- Ask the child to sort the pieces by shape.
- Have the child use a complete sentence to explain *sort.*

Inclusion Students

materials: Attribute Shape pieces (blue, yellow, and red) from Lesson 21

- Arrange three rows of Attribute Shapes; make the first row blue, the second row yellow, and the third row red. Keeping the shapes horizontally and vertically aligned, place the following shapes in each row in this order: large square, small square, large circle, large triangle.
- Ask the child to cover his or her eyes.
- Remove one of the shapes.
- Ask the child to uncover his or her eyes and to identify the missing shape.
- Ask the child how he or she knew which shape was missing.
- Replace the shape, and repeat with another shape piece.

At-Risk Students

- Math Center Activity 14

Objectives	Materials for Four Children	Teacher Directions
• Sorting Geometric Shapes by Attributes	• 4 bags of Attribute Shape pieces (from Lesson 21)	Pair the children. Ask one child to remove a piece from his/her partner's bag of shape pieces without letting the partner see. Ask the partner to sort his/her pieces to identify the missing attribute piece.

Advanced Learners

materials: Attribute Shape pieces (one set of each of the 4 colors) from Lesson 21 in a zip-top bag

- Ask the child to place the green shapes in a single line (in any order).
- Without the child seeing, remove three of the remaining shapes from the bag.
- Give the bag back to the child and ask him or her to find which shape pieces are missing.
- Repeat with four or more missing shape pieces, if desired.

- **Drawing Pictures and Writing Number Sentences for "Some, Some More" and "Some, Some Went Away" Stories**

English Learners

- Write 7 + 2 = 9 on the board.
- **number sentence:** a sentence that uses numbers and symbols but not words
- Ask the child to point to and say the numbers in the number sentence; ask him or her to point to and say the symbols.
- Write 7 flowers + 2 flowers = 9 flowers and note that number sentences may have labels.
- In class a label was used to show the items being talked about, such as flowers.
- Write the following in a column on the board: 8 pennies – 2 pennies = 6 pennies; 3 toys and 5 birds; 5 cats = 4 cats + 1 cat; 7 – 3 = 4.
- Ask the child which of these is not a number sentence. (3 toys and 5 birds)
- Have the child use a complete sentence to explain *number sentence.*

Inclusion Students

materials: 8 index cards; create four pairs of index cards: for each pair, on one card write a number sentence, such as 3 + 5 = 8, and on the other show a pictorial representation similar to those drawn for Lesson 22; make two addition pairs and two subtraction pairs

- Ask the child to match the card with the picture with its number sentence.

At-Risk Students

- Math Center Activity 15

Objectives	Materials for Four Children	Teacher Directions
• Writing "Some, Some More" and "Some, Some Went Away" Stories	• story paper—2 pieces per child	Ask each child to write a *some, some, more* story and a *some, some went away* story and draw pictures to illustrate his/her stories.

Advanced Learners

- Extend and Challenge CD, Activity 4, At the Amusement Park
- After the child finishes the activity, use the question(s) on the recording form to evaluate his or her understanding. (Monitoring Student Progress binder, Extend and Challenge CD booklet)

Guide to Differentiated Instruction

- **Dividing a Shape in Half**
- **Shading One Half of a Shape**
- **Identifying if a Fractional Part Of a Whole Is Closer to 0, 1/2, or 1**

English Learners

- Draw two equal-size squares on the board; divide the first square in half with a vertical line and shade half; in the second square draw an off-center line and shade one portion.
- **divide:** to separate into equal groups
- Explain to the child that the first square is divided in half, but the second square is not divided into equal parts and each part is not a half.
- Draw a circle and a rectangle on the board.
- Ask the child to divide each figure in half and to shade one half.
- Have the child use a complete sentence to explain *divide.*

Spanish Cognate: divide dividir

Inclusion Students

The child will identify if a fractional part of a whole is closer to 0, 1/2, or 1.

materials: color tiles (16 yellow and 16 green)

- Using the 16 yellow color tiles, make a 4-by-4 square, with all the squares touching.
- Point out to the child that the yellow square shows a whole.
- Replace one of the corner tiles with a green tile; explain that the yellow portion of the square is still close to a whole.
- Replace two more yellow tiles adjacent to the green corner with green tiles.
- Ask the child whether the yellow part of the square is closer to a whole or a half. (whole)
- Replace four more yellow tiles with green on the same side as the other green tiles.
- Ask whether the yellow part of the square is closer to a whole or a half. (half)
- Replace six more yellow tiles with green.
- Ask whether the yellow part of the square is closer to a whole, a half, or zero. (zero)
- Replace the remaining yellow tiles with a green; explain that the part of the square that is yellow is now zero.

At-Risk Students

- Math Center Activity 16

Objectives	Materials for Four Children	Teacher Directions
• Identifying the Missing Half or Third of a Shape	• 4 Learning Palettes • 4 Learning Palette Card 9	Ask each child to match the problems and answers on Learning Palette Card 9.

- **Dividing a Square in Half Two Different Ways**

English Learners

- Draw three equal-size squares. Divide the first two squares in half (one with a vertical line and one with a diagonal line). Shade one half of each square. In the third square, draw an off-center line.
- **halves:** two equal parts that make one whole; the plural of "half"
- Remind the child that since the last square is not divided into equal portions, it does not show an object divided in half.
- Explain that the first two squares are divided in half—with two equal parts called *halves.*
- Shade the other half of the first two squares, showing that each is made up of two equal parts.
- Have the child use a complete sentence to explain *halves.*

Inclusion Students

materials: square piece of paper

- Have the child fold the square in half one way.
- Then show the child another way to divide a square in half.
- Using the square piece of paper, ask the child to explain/demonstrate how he or she knows when something is divided in half. (The two parts cover each other exactly.)

At-Risk Students

materials: green and blue tangram sets

- Create a square using 1 large green and 1 large blue triangle.
- Create a second square using the smallest two blue triangles and the blue square to make one half and the smallest two green triangles and the green square to make the other half.
- Explain to the child that the two squares are equal in size, but divided in half in two different ways; both squares show one half blue and one half green.
- Ask the child to take the two large triangles that form the first square and make one larger triangle.
- Then ask him or her to use the two large triangles to make a parallelogram.

- **Addition Facts: Doubles Plus 1**

English Learners

materials: classroom number line

- **number line:** a straight path marked with numbers at equal spaces
- Point out that the classroom number line shows numbers that follow each other from left to right, each number increasing in value by 1.
- Ask the child to point to and say one number on the number line, and then say the numbers that are one more and one less than the first number.
- Ask the child how a number line is different from a hundred number chart.
- Have the child use a complete sentence to describe a *number line.*

Inclusion Students

materials: Teacher Fact Cards (Set C, Doubles + 1, up to 5 + 6 and 6 + 5)

- Review the strategy of finding the smaller number, doubling it, and then adding 1 to find the answer.
- Show the Teacher Fact Cards in random order; ask the child to say the answers.
- Review any facts the child misses.

At-Risk Students

materials: Teacher Fact Cards (Set C, Doubles + 1)

- Review the strategy of finding the smaller number, doubling it, and then adding 1 to find the answer.
- Show the Teacher Fact Cards in random order; ask the child to say the answers.
- Review any facts the child misses.

Advanced Learners

- Ask the child: When you add doubles, will the sum of the two numbers be even or odd?
- Add various doubles with the child to find out: 1 + 1, 5 + 5, 8 + 8, 7 + 7.
- Ask if it matters whether the number that is being doubled (the number you start with) is even or odd. (no) If you double an even number, the answer is even; if you double an odd number, the answer is even.
- Ask the child: When you add an even number to an odd number, will the sum be even or odd?
- Add various combinations with the child to find out: 2 + 3, 3 + 8, 1 + 6.
- Ask the child: If you have a doubles + 1 fact, will the answer be even or odd?
- Ask why. (You will always be adding an even number to an odd number. Since the doubles plus one numbers will always be two numbers that come one after another on the number line, one of the numbers will always be even and the other will always be odd. Another way to think about it is that adding doubles will always result in an even number; adding one more must therefore make it odd: 6 + 6 + 1)
- Show some examples that the sum of a doubles + 1 problem is odd: 4 + 5, 7 + 8, 11 + 12.

- **Identifying Geometric Shape Pieces That Differ in One Way**

English Learners

materials: Attribute Shape pieces (large red circle; small green triangle; and small blue square) from Lesson 21

- Place the three shape pieces in front of the child.
- **attribute:** a feature or characteristic of something or someone
- Explain that an attribute of something is a word that can be used to describe it.
- Point out the attributes of the circle: it is large, red, and round.
- Point out the attributes of the triangle: it is small, green, and has three sides and three corners.
- Ask the child to tell you three attributes, such as size, color, and number of sides or corners, of the square.
- Have the child use a complete sentence to explain *attribute*.

Spanish Cognate: attribute attributo

Inclusion Students

materials: Attribute Shape pieces from Lesson 21 (two colors)

- Ask the child to create a train of shape pieces, in which the pieces must match in two ways (color, size, or shape).

At-Risk Students

- Math Center Activity 17

Objectives	Materials for Four Children	Teacher Directions
• Identifying Geometric Shape Pieces That Differ by One Attribute	• 2 bags of Attribute Shape pieces (from Lesson 21)	Pair the children. Ask the children to work with a partner and play the Attribute Train Game introduced in Lesson 25-2.

- **Telling and Showing Time to the Half Hour**

English Learners

materials: demonstration clock and two student clocks

- **half hour:** a measurement of time equal to 30 minutes; one half of an hour
- Remind the child that in one hour, the minute hand moves around the clockface one time.
- On the demonstration clock, show 6:00; show the movement of the minute hand once around the clock until the clock shows 7:00.
- Review that in a half hour, the minute hand moves halfway around the clock.
- On the demonstration clock, show the movement of the minute hand from 6:00 to 6:30.
- Set one of the student clocks to 9:00; ask the child to set the other student clock to 9:30.
- Ask the child to write 9:00 and 9:30 digitally.
- Have the child use a complete sentence to explain *half hour.*

Inclusion Students

materials: student clock

- Write 11:30 on the board.
- Ask the child to tell you what time it is and to set the clock to match.
- Repeat with the following digital times: 2:30, 5:30, 8:30.

At-Risk Students

- Math Center Activity 18

Objectives	Materials for Four Children	Teacher Directions
• Telling Time to the Half Hour	• 4 Learning Palettes • 4 Learning Palette Card 10	Ask each child to match the problems and answers on Learning Palette Card 10.

Advanced Learners

Journal Writing (Overview 3)

- Ask the child to name his or her favorite television show.
- Have the child write: what time his or her favorite television show comes on; what time the show ends; how long the show lasts.

- **Estimating Temperature**
- **Reading a Thermometer to the Nearest 10 Degrees**

English Learners

materials: Meeting Board thermometer and Indoor/Outdoor thermometer

- **temperature:** the measure by degree of how hot or cold something is
- **thermometer:** a tool used to measure temperature
- Using both thermometers, review the parts of the thermometer with the child—the lines and numbers that show the temperature in degrees, the scale indicator (F and C), and the column of liquid.
- Remind the child that the liquid goes up the glass as the temperature goes up.
- Ask what he or she expects the liquid in the thermometer to show when its cold.
- Have the child use a complete sentence to explain *temperature* and *thermometer*.

Spanish Cognates: temperature temperature thermometer termómetro

Inclusion Students

The child will estimate temperature.

materials: Meeting Board thermometer and Indoor/Outdoor thermometer

- Using both thermometers, review the parts of the thermometer with the child—the lines and numbers that show the temperature in degrees, the scale indicator (F and C), and the column of liquid.
- Shade the thermometer to 58°.
- Ask the child *about* what temperature is shown on the thermometer.
- Repeat with the following temperatures: 41°, 8°, 29°.

At-Risk Students

The child will read a thermometer to the nearest 10 degrees.

materials: Meeting Board thermometer

- Shade the thermometer to 54°.
- Ask the child if 54 is closer to 50 or 60.
- Have the child write the temperature to the nearest 10 degrees on the board.
- Repeat with the following temperatures: 42°, 68°, 66°, 29°, 32°.

Advanced Learners

- Extension and Enrichment problem for Guided Class Practice 27, problem 5 (Extension and Enrichment Student Problem [page 5] at www.SaxonMathResources.com)

Extension and Enrichment Problem	Answer
Draw a clock face with numbers. Draw the hour and minute hands to show the time 2 hours later than 8:30. Show the new time on a digital clock too.	analog and digital clocks showing 10:30

• **Counting Dimes and Pennies**

English Learners

materials: 10 pennies and 1 dime

- **penny:** a U.S. coin worth one cent
- **dime:** a U.S. coin worth ten cents
- Place the coins in front of the child and review that ten pennies is the same amount of money as one dime.
- Have the child point out the similarities and differences in the two types of coins (color, size, what appears on the coins, etc.); note that *one cent* appears on the penny and *one dime* appears on the dime.
- Ask the child which coin is worth more; review the value of both coins.
- Have the child use a complete sentence to describe both a *penny* and a *dime.*

Spanish Cognate: cent centavo

Inclusion Students

materials: Lesson Workmat 28 and one container of dimes and pennies (fewer than ten of each)

- Ask the child to sort the coins on the Workmat and tell you how many of each there are.
- Ask him or her to tell you the total amount of money.
- Repeat with other combinations of dimes and pennies, as desired.

At-Risk Students

- Math Center Activity 20

Objectives	Materials for Four Children	Teacher Directions
• Counting Dimes and Pennies	• 4 Learning Palettes • 4 Learning Palette Card 11	Ask each child to match the problems and answers on Learning Palette Card 11.

Advanced Learners

- Extension and Enrichment problem for Guided Class Practice 28, problem 3 (Extension and Enrichment Student Problem [page 5] at www.SaxonMathResources.com)

Extension and Enrichment Problem	Answer
Katie had 2 dimes and 1 penny. Mallory gave her 3 more dimes. Draw a picture to show all the coins she has now. What is the total value of all her coins?	a picture showing 5 dimes and 1 penny; 51¢

- **Writing Addition and Subtraction Fact Families**

English Learners

- On the board write: $3 + 4 = 7$, $4 + 3 = 7$, $7 - 3 = 4$, $7 - 4 = 3$.
- **fact family:** a group of addition and subtraction facts that all use the same three numbers
- Remind the child that a fact family with three different numbers will always have two addition and two subtraction facts, as shown in the example on the board with the numbers 3, 4, and 7.
- Write 1, 5, and 6 on the board.
- Ask the child to write the fact family, two addition and two subtraction facts, using these three numbers.
- Have the child use a complete sentence to explain *fact family*.

Spanish Cognate: family familia

Inclusion Students

materials: 12 index cards; create three sets of fact families, as shown, with one fact on each index card:

Set 1: $6 + 2 = 8$, $2 + 6 = 8$, $8 - 6 = 2$, $8 - 2 = 6$
Set 2: $3 + 4 = 7$, $4 + 3 = 7$, $7 - 3 = 4$, $7 - 4 = 3$
Set 3: $5 + 1 = 6$, $1 + 5 = 6$, $6 - 1 = 5$, $6 - 5 = 1$

- Ask the child to sort the cards into groups that are fact families.

At-Risk Students

- Math Center Activity 21

Objectives	Materials for Four Children	Teacher Directions
• Identifying the Missing Addition or Subtraction Fact in a Fact Family	• 4 Learning Palettes • 4 Learning Palette Card 12	Ask each child to match the problems and answers on Learning Palette Card 12.

Advanced Learners

- On the board write 7, 8, and 15.
- Ask the child to write the four facts that make up a fact family.
- On the board, write 9, 9, and 18.
- Ask the child to write the facts for these numbers and ask how many different facts there are. (If necessary, remind him or her that doubles have only two facts in their fact family.)

- **Addition Facts: Sums of 8 and 9**
- **Drawing a Picture to Solve a Problem**
- **Looking for a Pattern to Solve a Problem**

English Learners

materials: Chart 23 (Problem-Solving Process chart)

- **strategy:** a plan of action or a method to solve a problem
- Under the *Plan* section of Chart 23, review with the child types of strategies that may be used in problem-solving. (Those that have been used so far are: Act It Out, Use Logical Reasoning, Draw a Picture, and Look for a Pattern.)
- Ask the child to name his or her favorite problem-solving strategy and explain why.
- Have the child use a complete sentence to explain *strategy*.

Spanish Cognate: strategy estrategia

Inclusion Students

The child will practice addition facts for sums of 8 and 9.

materials: color tiles and Teacher Fact Cards (Set D, part of Set A [2 + 6, 2 + 7, 6 + 2, 7 + 2], B [0 + 9], and C [4 + 5, 5 + 4])

- Show the Teacher Fact Cards in random order; ask the child to say the answers.
  Note: If the child has difficulty, allow him or her to use the color tiles to act out the problem.

At-Risk Students

materials: Problem-Solving Worksheet 30A and Performance Task Worksheet 30B

- Review with the child the problem-solving strategies and Problem-Solving Worksheet 30A and Performance Task Worksheet 30B.

Advanced Learners

- Extension and Enrichment problem for Problem-Solving Worksheet 30A (Extension and Enrichment Student Problem [page 6] at www.SaxonMathResources.com)

Extension and Enrichment Problem	Answer
The boxcars on a toy train are in a blue, yellow, red, yellow, green repeating pattern. Show the color of the 14th boxcar. There are a total of 18 cars on the toy train. If the train were half as long, how many boxcars would be on the train? Draw a picture of the shorter train. What color would the last boxcar be?	yellow; 9 boxcars; a picture showing 9 colored boxcars; yellow

- **Identifying Geometric Shape Pieces That Are Alike in Only One Way**

English Learners

- **rule:** direction or instruction
- Explain that when games are played, a set of rules is followed so that the game is fair to every player.
- Remind the child that in the classroom, rules are put into place so that each child understands what is expected of him or her.
- Ask the child to name one or two of your classroom rules. (no chewing gum, no talking out of turn, etc.)
- Have the child use a complete sentence to explain *rule*.

Inclusion Students

materials: Attribute Shape pieces from Lesson 21 (two colors)

- Ask the child to create a train of pieces in which the pieces that follow match in only one way (color, shape, or size).

At-Risk Students

- Math Center Activity 22

Objectives	Materials for Four Children	Teacher Directions
• Identifying Geometric Shape Pieces That Are Alike in Only One Way	• 2 bags of Attribute Shape pieces (from Lesson 21)	Pair the children. Ask the children to work with a partner and play the Attribute Train Game introduced in Lesson 30-2.

Advanced Learners

Journal Writing (Overview 3)

materials: pattern blocks

- Have the child choose 2 different pattern blocks and then trace them.
- Ask the child to identify the pattern blocks and describe two ways they are the same and two ways they are different.

- **Creating and Reading a Bar Graph**

English Learners

materials: the child's Lesson Worksheet 31 (Weekday Wake-Up Times)

- Show Lesson Worksheet 31 to the child.
- **bar graph:** a graph that shows information using parallel rows or columns
- Remind the child that a graph is a way to show information.
- Using the child's Lesson Worksheet, point to the columns representing the wake-up times, and the height of each column, which indicates the number of children with that wake-up time.
- Ask the child what time the smallest number of children wake up.
- Ask him or her what is the latest time someone in the class wakes up.
- Have the child use a complete sentence to describe *bar graph.*

Spanish Cognate: bar graph gráfica de barras

Inclusion Students

materials: Lesson Worksheet 31 (Weekday Wake-Up Times)

- Using the graph, ask the child questions related to the graph. (Examples: How many children wake up at 8:00? At what time do the least number of children wake up? How many children in all wake up at the same time you do?)

At-Risk Students

materials: Lesson Worksheet 31 (bar graph, Weekday Wake-Up Times) and Lesson Worksheet 17 (pictograph, Favorite Day of the Week)

- Ask the child how a bar graph differs from a pictograph.

Advanced Learners

Journal Writing (Overview 4)

materials: Lesson Worksheet 31 Chart

- Ask the child to write three things he or she learned from the wake-up graph.

Guide to Differentiated Instruction Lesson 32

- **Tallying**
- **Counting by 5's**

English Learners

materials: color tiles (22 red and 18 yellow)

- **tally:** to keep track while counting by writing small vertical and oblique marks
- On a sheet of paper, make two columns with the headings *red* and *yellow.*
- Draw five red tiles from the stack and review with the child how to make tally marks.
- Ask the child to sort the remaining tiles by color and to tally them.
- Have the child use a complete sentence to explain *tally.*

Inclusion Students

The child will count by 5's.

materials: Math Folder (hundred number chart)

- Have the child count by 5's to 100.

At-Risk Students

- Math Center Activity 25

Objectives	Materials for Four Children	Teacher Directions
• Counting Tally Marks	• 4 Learning Palettes • 4 Learning Palette Card 13	Ask each child to match the problems and answers on Learning Palette Card 13.

- Math Center Activity 26

Objectives	Materials for Four Children	Teacher Directions
• Counting by 5s	• 4 Learning Palettes • 4 Learning Palette Card 14	Ask each child to match the problems and answers on Learning Palette Card 14.

Advanced Learners

- Online Activity, Tally Marks
  (Student Activity Center is available at www.SaxonMathActivities.com.)

• Identifying Horizontal, Vertical, and Oblique Lines

English Learners

materials: Math Folder (Line Segments), index card with a large right triangle drawn, and three crayons (blue, red, yellow)

- Show and review the three types of line segments shown in the Math Folder.
- **horizontal:** a path that goes from left to right
- **vertical:** a path that goes up and down
- **oblique:** a straight path that is slanted; a line neither horizontal nor vertical
- Have the child trace the line segments of the right triangle as you give the following directions: trace the horizontal line segment with the blue crayon, trace the vertical line segment with the red crayon, trace the oblique line segment with the yellow crayon.
- Allow the child to refer to the Math Folder as he or she completes the activity.
- Have the child use complete sentences to explain *horizontal, vertical,* and *oblique.*

Spanish Cognates:

horizontal	horizontal
vertical	vertical
oblique	oblicuo(a)

Inclusion Students

materials: Math Folder (Line Segment)

- Review the three types of line segments shown in the Math Folder.
- Ask the child to use: his or her right arm to make an oblique line, his or her left arm to make a horizontal line, and his or her left arm to make a vertical line.
- Allow the child to refer to the Math Folder as he or she completes the activity.

At-Risk Students

- Math Center Activity 27

Objectives	Materials for Four Children	Teacher Directions
• Assembling a Hundred Number Chart	• 4 sets of hundred number chart pieces (from Lesson 33)	Ask each child to assemble the pieces to make a hundred number chart.

Advanced Learners

- Extension and Enrichment problem for Guided Class Practice 33, problem 2 (Extension and Enrichment Student Problem [page 7] at www.SaxonMathResources.com)

Extension and Enrichment Problem	Answer
Draw a triangle. Can you draw a triangle using only horizontal and vertical line segments? Draw a square. Can you draw a square using line segments that are not horizontal and vertical? If so, draw one.	no; yes

Guide to Differentiated Instruction

- **Dividing a Whole into Halves, Fourths, and Eighths**
- **Comparing Halves, Fourths, and Eighths**
- **Writing a Unit Fraction Using Fraction Notation**

English Learners

- **one fourth:** 1/4; one of four equal parts
- **one eighth:** 1/8; one of eight equal parts
- Draw three circles on the board; under the first write *halves*, under the second write *fourths*, and under the third write *eighths*.
- As you describe your actions, divide the first circle in half and label each half with the fraction 1/2.
- Ask the child how many halves make up the circle. (2)
- Repeat with fourths and eighths.
- Explain to the child that fractions and ordinal numbers may use the same term (such as fourth and eighth), but have different meanings; give an example, such as, fourth could mean a person who is fourth in line.
- Have the child use complete sentences to explain *one fourth* and *one eighth.*

Inclusion Students

The child will review whole, halves, and fourths.

materials: color tiles (7 green and 5 blue)

- Create one square with 4 green tiles.
- Create a second square with 2 green and 2 blue tiles.
- Create a third square with 1 green and 3 blue tiles.
- Review with the child that the completely green square represents one whole.
- Ask the child to describe how the second square is divided, and what each color represents. (one half)
- Ask the child what part of the last square is green. (If necessary have the child count the total number of squares making up the last square, and then say what part of that total is green.)

At-Risk Students

The child will review halves, fourths, and eighths.

materials: color tiles (7 green and 17 blue)

- Create a rectangle with 4 blue and 4 green tiles.
- Create a second rectangle with 6 blue and 2 green tiles.
- Create a third rectangle with 7 blue and 1 green tiles.
- Review with the child that a completely green rectangle would represent one whole.
- Ask the child to describe how the first rectangle is divided, and what each color represents. (one half)
- Ask the child to compare the green portion of the second and third rectangles and tell you what fraction each represents.

• Addition Facts: Sums of 10

English Learners

materials: playing cards (A, 2, 3, 4, two 5's, 6, 7, 8, 9)

- **concentration:** focus thinking and attention on something
- Explain that when you concentrate on something, you are paying attention to it and focused on it, without being distracted or having your mind wander onto something else.
- Place the cards face down in random order in two rows and five columns.
- Ask the child to turn over two cards at a time. If the sum of the cards is 10, the child will pick up the two cards and take another turn. If the sum is not 10, you take a turn. Repeat until all the cards are removed. (Use the Ace card as a one.)
- Ask the child to explain why the game just played is called *Concentration.*

Spanish Cognate: concentration concentración

Inclusion Students

materials: playing cards (A–9 of clubs and hearts)

- Ask the child to place the clubs in a row in order from A to 9.
- Remind the child that the ace represents one.
- Ask the child to place a heart card on top of the club that will result in a sum of 10.
- Repeat, placing the clubs in a row in random order.

At-Risk Students

- Math Center Activity 28

Objectives	Materials for Four Children	Teacher Directions
• Identifying Sums of 10	• 4 decks of playing cards with the face cards removed	Allow the children to play the Making 10 Game introduced in Lesson 35-1.

Advanced Learners

- Learning Palette Enrichment Card E3, Identifying a Missing Addend

- **Weighing Objects Using Nonstandard Units**
- **Comparing and Ordering Objects by Weight**

English Learners

materials: balance, color tile, crayon, and paper clip

- **lightest:** having less weight than all the others
- **heaviest:** having more weight than all the others
- Have the child place the color tile in one bucket of the balance and the paper clip in the other.
- Have him or her note which object weighs more, the color tile or the paper clip.
- Ask the child to replace the paper clip with the crayon and repeat the activity.
- Ask him or her which of the three objects weighs the most and which weighs the least; ask him or her to place the objects in order from heaviest to lightest.
- Have the child use complete sentences to explain *lightest* and *heaviest.*

Inclusion Students

The child will compare and order objects by weight.

materials: balance, color tiles, and three classroom objects, each with a weight equal to an exact number of color tiles

- Ask the child to weigh one of the objects using the color tiles.
- Ask him or her to remove the object and color tiles from the balance and place them next to each other on the table.
- Repeat with the remaining two objects.
- Ask the child to compare the number of color tiles needed for each object.
- Ask him or her which of the three objects weighs the most and which weighs the least, and to place the objects in order from heaviest to lightest.

At-Risk Students

- Math Center Activity 29

Objectives	Materials for Four Children	Teacher Directions
• Weighing Objects Using Nonstandard Units	• 5 classroom objects, each with a weight equal to an exact number (3–20) of color tiles • balance • container of color tiles • Center 29 Master *Make 1 copy for each child in class.*	Pair the children. Ask each pair to weigh an object using color tiles and write the name of the object and the number of color tiles used on Center 29 Master. Ask the children to repeat with the other four objects. Ask each pair to identify lightest and heaviest objects.

- **Adding 10 to a Multiple of 10**
- **Finding Missing Numbers on a Piece of the Hundred Number Chart**

English Learners

materials: Math Folder (hundred number chart)

- Draw a sketch or stick figure of a dog with three legs.
- **missing:** not present, not found in the usual place
- Point out that a dog normally has four legs but in the drawing one of the legs is missing.
- Draw a face without a nose and ask the child if he or she can see what is missing.
- Draw a 3 × 3 grid and put 65 in the middle square. Ask the child to fill in the missing numbers in the remaining squares, using the Math Folder hundred number chart for reference, as needed.
- Have the child use a complete sentence to explain *missing*.

Inclusion Students

The child will find missing numbers on the hundred number chart.

materials: Math Folder (hundred number chart) and color tiles

- Cover several numbers on the hundred number chart with the color tiles.
- Ask the child to identify the missing numbers.

At-Risk Students

- Math Center Activity 30

Objectives	Materials for Four Children	Teacher Directions
• Identifying the Number of Tens and Ones in a Number	• 4 Learning Palettes • 4 Learning Palette Card 15	Ask each child to match the problems and answers on Learning Palette Card 15.

Advanced Learners

Journal Writing (Overview 4)

materials: Math Folder (hundred number chart)

- Ask the child to choose a number on the hundred number chart.
- Have him or her write 3 clues to help another child guess the number.

- **Identifying Pairs**
- **Dividing a Set of Objects Into Groups of Two**

English Learners

- Draw a pair of socks on the board.
- **pair:** two items that are usually together such as shoes or gloves
- Point out several examples of pairs, such as earrings or shoes.
- Ask the child to point to and say what he or she has pairs of on his or her body. (eyes, ears, hands, feet, legs, arms, etc.)
- Have the child use a complete sentence to explain *pair.*

Spanish Cognate: pair par

Inclusion Students

The child will identify pairs and divide a set of objects into groups of two.

materials: 2 of each type of pattern blocks

- Ask the child to make pairs of matching pattern blocks.
- Ask him or her the total number of pattern blocks.
- Ask him or her the number of pairs of pattern blocks.

At-Risk Students

The child will identify pairs.

materials: paper divided into three columns labeled 1 Pair, 2 Pairs, 3 Pairs

- Ask the child to draw the number of pairs of socks indicated in each column.
- Ask him or her to circle each pair of socks.
- Have the child tell you the total number of socks and the total number of pairs of socks on the page. (12 socks and 6 pairs)

Advanced Learners

- Extension and Enrichment problem for Guided Class Practice 37, problem 2 (Extension and Enrichment Problem [page 7] at www.SaxonMathResources.com)

Extension and Enrichment Problem	Answer
How many socks would you have if you had 9 pairs and 1 white sock left over? How many socks would you have altogether if your brother gave you 3 more pairs of socks?	19 socks; 25 socks

- **Identifying Tens and Ones**

English Learners

materials: 10 pennies and 1 dime

- Show the child the pennies and dime.
- Review **penny:** a U.S. coin worth one cent
- Review **dime:** a U.S. coin worth ten cents
- Place the coins in front of the child and review that ten pennies is the same amount of money as one dime.
- Ask the child to point out similarities and differences between a penny and a dime.
- Have the child use a complete sentence to discuss *penny* and *dime*.

Inclusion Students

materials: Math Folder (hundred number chart)

- On the board draw two columns, with the titles *tens* and *ones* (similar to the columns used in the lesson).
- Point to a random number on the hundred number chart and ask the child to tell you the number of tens and number of ones in that number.
- Record these in the correct column.
- Repeat with additional numbers.

At-Risk Students

materials: dimes, pennies, and 5 index cards with money amounts (Example: 6 dimes, 4 pennies).

- Have the child select a card, show the total amount with coins, and write the amount using the cent symbol.
- Repeat with the other money amounts.

Advanced Learners

- Learning Palette Enrichment Card E4, Counting Sets of 10s and 1s

- **Identifying Halves, Fourths, and Eighths of a Whole**
- **Creating and Reading a Bar Graph**

English Learners

materials: book

- Draw a sketch of a book on the board.
- **solid:** an object that can be touched
- Show the child the real book and point out that the drawing is only a representation of the real item.
- Explain that the book is a solid object because it can be touched, unlike the drawing on the board.
- Ask the child what solid was used in class and divided into halves and fourths. (apple)
- Ask the child to point out another solid object and explain why it is a solid.
- Have the child use a complete sentence to explain *solid.*

Spanish Cognate: solid sólido

Inclusion Students

The child will identify halves and fourths of a whole.

materials: 6 index cards; on 4 cards show the following: a circle divided into halves, a circle divided into fourths, a square divided into halves, a square divided into fourths; on the remaining 2 cards, write *halves* and *fourths*

- Place the index cards with the words on a table in front of the child.
- Ask the child to match the appropriate picture cards to the words; help the child as needed.

At-Risk Students

- Math Center Activity 31

Objectives	Materials for Four Children	Teacher Directions
• Reading a Bar Graph • Identifying How Many More	• 4 Learning Palettes • 4 Learning Palette Card 16	Ask each child to match the problems and answers on Learning Palette Card 16.

Advanced Learners

Journal Writing (Overview 4)

materials: Our Favorite Apple graph

- Have the child write 3 questions he or she could ask the class about the apple graph.

- **Addition Facts: Sums of 11**
- **Making an Organized List to Solve a Problem**
- **Solving a Problem by Acting It Out**

English Learners

materials: shelves showing classroom books and manipulatives, and index cards with the letters: A, B, C, D, E, F, G, I, J

- Show the child that the shelves with the classroom books and manipulatives are kept in an organized way, so that it is easier to find the items when they are needed.
- **organize:** arrange in an orderly way
- Explain that solving a mathematics problem is easier when the individual parts of the problem are organized and better understood.
- Give the child the index cards with the alphabet letters.
- Ask him to tell you what letter is missing from the letters A through J.
- Explain that the child will need to organize the letters in alphabetical order to find the missing letter.
- Have the child use a complete sentence to explain *organize.*

Spanish Cognate: organize organizar

Inclusion Students

The child will practice the sums of 11.

materials: playing cards (A–10 of clubs and hearts)

- Ask the child to place the clubs in a row in order from A to 10.
- Remind the child that the ace represents one.
- Ask the child to place a heart card on top of the club that will result in a sum of 11.
- Repeat, placing the clubs in a row in random order.

At-Risk Students

materials: Problem-Solving Worksheet 40A and Performance Task Worksheet 40B

- Review with the child the problem-solving strategies and Problem-Solving Worksheet 40A and Performance Task Worksheet 40B.

Advanced Learners

- Extension and Enrichment problem for Problem-Solving Worksheet 40A (Extension and Enrichment Student Problem [page 8] at www.SaxonMathResources.com)

Extension and Enrichment Problem	Answer
Blake has a set of blocks made out of different shapes: squares, triangles, circles, rectangles, and hexagons. Show all the ways he can combine them in pairs if each pair is made up of 2 different shapes. How many different ways can Blake combine his blocks this way?	ST, SC, SR, SH, TC, TR, TH, CR, CH, and RH; 10 ways

- **Measuring With One-Inch Color Tiles**

English Learners

materials: color tiles and two small classroom objects

- Remind the child that a color tile is one inch long.
- **inch:** a small U.S. Customary System unit used to measure length, height, or distance; in.
- Have the child observe as you measure a small classroom object with the color tiles; say the object's length in tiles and inches.
- Have the child measure the second object and tell you its length in tiles and inches.
- Have the child use a complete sentence to explain *inch.*

Inclusion Students

materials: color tiles and several small classroom objects (pencil, crayon, Math Folder, eraser, etc.)

- Remind the child that each color tile is 1 inch long.
- Ask the child to use the color tiles to measure the length of the classroom objects; measure the Math Folder from top to bottom.
- Ask him or her to tell you the lengths of the objects in tiles and inches.

At-Risk Students

- Math Center Activity 33

Objectives	Materials for Four Children	Teacher Directions
• Measuring Length Using Nonstandard Units	• 10–12 classroom objects of varying lengths • container of 1" color tiles • Center 33 Master *Make 1 copy for each child in class.*	Pair the children. Ask each pair to measure the lengths of five objects using color tiles and write the names of the objects and the number of color tiles used on Center 33 Master.

Advanced Learners

materials: ruler and 5 straws cut to various whole-inch lengths

- Review with the child that each large, numbered segment on the ruler is one inch.
- Ask the child to sort the straws from shortest to longest.
- Ask the child to measure and record the length of each straw.

- **Naming Fractional Parts of a Whole**
- **Comparing Fractional Parts of a Whole**
- **Writing a Fraction Using Fraction Notation**

English Learners

- On the board draw three circles and label them 1/4, 3/8, and 1/2; divide the first circle into fourths with 1/4 shaded, the second circle into eighths with 3/8 shaded, and the third circle into halves with 1/2 shaded.
- **fraction:** a number that tells the number of parts (numerator) of a whole (denominator)
- Point to each fraction and remind the child that the bottom number in a fraction shows how many parts make up the whole.
- Review that the top number in a fraction tells the number of parts of the whole being discussed.
- Draw a circle divided into sixths and shade 2/6.
- Have the child write the fraction.
- Write the fraction 2/3 on the board.
- Draw a circle divided into thirds and have the child shade 2/3 of it.
- Have the child use a complete sentence to explain *fraction.*

Spanish Cognate: fraction fracción

Inclusion Students

The child will compare and match fractional parts of a whole.

materials: index cards showing fractions of circles shaded (1/2, 1/4, 1/6), fractions of squares shaded (1/2, 1/4), and index cards showing the fractions 1/2, 1/4, and 1/6

- Have the child match the index cards showing the fractions with the cards showing the shaded figures.

At-Risk Students

The child will compare fractional parts of a whole.

materials: crayons (blue, red, green) and 3 index cards; on each index card draw a circle, one divided into halves, one into sixths, and one into twelfths

- Ask the child to color 1/2 of the first circle blue, 3/6 of the next circle red, and 6/12 of the last circle green.
- Ask him or her to compare the colored portions and tell you what he or she notices.

Advanced Learners

Journal Writing (Overview 5)

- Ask the child to say whether he or she would rather have 1/2 of a pizza or 1/6 of a pizza, and write a paragraph and draw a picture to explain why.

• Trading Pennies for Dimes

English Learners

materials: 17 color tiles

- Make three stacks of color tiles; in the first put 6 tiles, in the second put 9 tiles, and in the last put 2 tiles.
- **fewest:** the least number of all
- Review with the child that when comparing two things, the word used to compare often ends in *er,* and when you compare three or more things, the comparing word often ends in *est.*
- Show him or her that if you compare the number of tiles in the first two stacks, the first has *fewer* than the second. If you compare the number of tiles in all three stacks, the last has the *fewest* or lowest number.
- Have the child use a complete sentence to explain *fewest.*

Inclusion Students

materials: cups of 25 pennies, 31 pennies, and 5 dimes

- Give the child the first cup of pennies and ask him or her to put as many pennies as possible into groups of 10.
- Ask him or her to trade each group of 10 pennies for 1 dime.
- Ask the child to tell you how many dimes and pennies there are, and what the total money amount equals.
- Repeat with the cup of 31 pennies.

At-Risk Students

- Math Center Activity 35

Objectives	Materials for Four Children	Teacher Directions
• Trading Pennies and Dimes	• 4 Learning Palettes • 4 Learning Palette Card 17	Ask each child to match the problems and answers on Learning Palette Card 17.

Advanced Learners

- Extension and Enrichment problem for Guided Class Practice 42, problem 2 (Extension and Enrichment Student Problem [page 9] www.SaxonMathResources.com)

Extension and Enrichment Problem	Answer
Draw 3 squares of the same size. Divide the first square into fourths and shade two fourths. Divide the second square into eighths and shade four eighths. Divide the last square into thirds and shade two thirds. Write the fraction you shaded next to each square. Which square shows the greatest shaded amount? Are any of the shaded amounts equal? If so, which ones?	2/4, 4/8, 2/3; the last square (2/3); yes, the first and second squares each show one half (1/2) shaded

• Measuring to the Nearest Inch

English Learners

materials: ruler and two small classroom objects

- Give the child the ruler; review the inch markings and the 0-mark that shows where to begin measurement.
- **length:** the distance from one end to the other end
- Review with the child that length is the measure of how long an object is; point out the similarity of the words long and length.
- Working with the child, measure the length of one of the objects in inches.
- Ask the child to find the length of the second object in inches.
- Have the child use a complete sentence to explain *length*.

Inclusion Students

materials: ruler and 5 straws cut into various whole-inch lengths

- Give the child the ruler; review the inch markings and the 0-mark that shows where to begin measurement.
- Ask the child to sort the straws from longest to shortest.
- Ask him or her to measure each straw.
- Ask him or her to tell you the length of each straw in inches.

At-Risk Students

materials: ruler and 5 straws cut into various whole-inch lengths

- Give the child the ruler; review the inch markings and the 0-mark that shows where to begin measurement.
- Ask the child to sort the straws from longest to shortest.
- Ask him or her to measure each straw and record the length in inches.
- Remind the child to include the unit (inches) when recording the length.

Advanced Learners

Journal Writing (Overview 5)

- Ask the child to write the name of something that is about 6 inches long and to write clues to help others guess the object.

- **Adding 10 to a Two-Digit Number**

English Learners

materials: Math Folder (hundred number chart)

- Ask the child to use the hundred number chart for reference as you review the terms.
- **two-digit:** a number that is greater than 9 and less than 100; a number with two places
- **ones' digit:** the number in the ones' (or pennies) place
- **tens' digit:** the number in the tens' (or dimes) place
- Point to the following numbers on the hundred number chart; ask the child to read the number and to tell you the ones' and tens' digits: 50, 26, 41, 19.
- Have the child use complete sentences to explain *two-digit, ones' digit,* and *tens' digit.*

Inclusion Students

materials: 5 index cards with addition problems showing adding 10 a two-digit number

- Have the child answer with the correct sums.

At-Risk Students

- Math Center Activity 36

Objectives	Materials for Four Children	Teacher Directions
• Adding 10 to a Number Using Mental Computation	• 4 Learning Palettes • 4 Learning Palette Card 18	Ask each child to match the problems and answers on Learning Palette Card 18.

Advanced Learners

materials: Math Folder (hundred number chart)

- On the board write 18 + 20 = .
- Ask the child to find the sum, using the hundred number chart for reference, if needed.
- Ask him or her to explain what happens when 20 is added to a two-digit number. (the tens' digit becomes 2 more and the ones' digit stays the same)
- Ask the child to find the sum of each of the following number sentences: 48 + 20 =, 20 + 32 =, 36 + 30 =.

• Addition Facts: Sums of 12

English Learners

materials: 2 index cards; on one write 4 + 1 = 5 and on the other write the same numbers as a vertical problem

- Show the child the fact cards.
- Review **addend:** one of the numbers being combined in an addition problem to form the sum
- Review **sum:** the answer to an addition problem
- Explain to the child that *sum* and *some* are pronounced the same way, but have different meanings; write both words on the board and explain their meanings.
- Point out the addends and the sum on each card; show that the direction of the problem does not change the name of the terms.
- Have the child use complete sentences to explain *addend* and *sum.*

Inclusion Students

The child will practice the sums of 12.

materials: playing cards (2–10 of clubs and hearts)

- Ask the child to place the clubs in a row in order from 2 to 10.
- Ask him or her to place a heart card on top of the club that will result in a sum of 12.
- Repeat, placing the clubs in a row in random order.

At-Risk Students

- Math Center Activity 37

Objectives	Materials for Four Children	Teacher Directions
• Identifying Sums of 12	• 4 decks of playing cards with the face cards removed	Allow the children to play the Making 12 Game introduced in Lesson 45-1.

- **Identifying 1-Cup and 1/2-Cup Measuring Cups, Tablespoons, Teaspoons, and 1/2 Teaspoons**
- **Reading a Recipe**

English Learners

materials: recipe from Chart 45-2 and a cookbook (if available)

- Show the child the recipe from the chart and one from the cookbook.
- **recipe:** the list of ingredients and directions for making something
- On the recipe point out the list of ingredients used and the instructions for mixing and baking.
- Mention that when following a recipe, it is important to make sure that all of the ingredients are available by assembling them before beginning the recipe.
- Review that recipes are useful so that the food is prepared correctly and will be the same each time it is made.
- Ask the child where recipes can be found. (cookbooks, on the backs of food packages, newspapers, televisions cooking shows, online, etc.)
- Have the child use a complete sentence to explain *recipe.*

Inclusion Students

materials: dry measuring cup, half cup, and quarter cup, tablespoon, teaspoon, and half teaspoon

- Ask the child to order the six measuring tools from largest to smallest.
- Ask him or her to identify the following: half cup, teaspoon, half teaspoon, cup, quarter cup, tablespoon. (Show that the measuring tools have the size printed on them.)

At-Risk Students

materials: all measuring cups and spoons

- Review the differences between the liquid and dry measuring cups, then remove the liquid measuring cup from the table.
- Place the remaining measuring cups and spoons on the table in front of the child.
- Ask him or her to order them from largest to smallest.
- Ask the child to identify the tools; explain that their sizes are printed on them.
- Review that teaspoon may be abbreviated *t* or *tsp* and that tablespoon may be abbreviated *T* or *Tbsp;* the child can use this hint to remember that the small t means the smaller spoon and the capital T is the larger spoon.

Advanced Learners

Journal Writing (Overview 5)

- Ask the child what his or her favorite food is.
- Ask him or her to write a recipe for this favorite food.

- **Identifying Similarities and Differences Among Coins**
- **Counting Nickels**

English Learners

materials: penny, nickel, and dime

- Show the child the penny, nickel, and dime.
- **nickel:** a U.S. coin worth five cents
- Ask the child to examine the nickel; note that the words *five cents* appear on the back of the coin.
- Ask the child to tell you some similarities and differences among the three coins.
- Ask him or her how much each coin is worth, how many pennies can traded for a nickel, and how many nickels can traded for a dime.
- Have the child use a complete sentence to describe a *nickel.*

Inclusion Students

The child will review counting nickels.

materials: a cup of 11 nickels and 2 classroom items with the following price tags: 20¢ and 35¢

- Ask the child to pay for the 20¢ item using nickels.
- Have the child count by 5's as he or she points to each nickel and finds how much the total is.
- Repeat for the 35¢ item.

At-Risk Students

- Math Center Activity 38

Objectives	Materials for Four Children	Teacher Directions
• Showing Money Amounts Using Dimes and Nickels	• 4 classroom items with the following price tags: 20¢, 25¢, 30¢, and 35¢ • 4 cups of 10 dimes and 10 nickels each • Center 38 Master *Make 1 copy for each child in class.*	Ask each child to select an item with a price tag and use the coins to find three different ways to pay for the item. Ask the child to record the combinations of coins used on Center 38 Master.

Advanced Learners

materials: 2 or 3 different nickels (for example: pre-2004, Peace [Louisiana Purchase], Keelboat [Lewis and Clark], Buffalo, Ocean View, or new Jefferson face)

- Ask the child to describe or write about the images on the different nickels.

• Writing the Date Using Digits

English Learners

materials: the Meeting Board Today's date strip

- Point to the strip that says *Today's date is* __________ on the Meeting Board.
- **date:** the month, day, and year
- Review with the child that when dates are written in digits, it is the custom in the United States to write the number of the month first, followed by the number of the day, and then the year.
- Have the child write the following dates using digits separated by slashes: January 10, 1998; March 11, 2005; and June 2, 2004.
- Have the child use a complete sentence to explain *date.*

Inclusion Students

materials: 5 pairs of index cards; for each pair, write one with a date using words and the other with the same date using digits.

- Ask the child to match the cards.

At-Risk Students

materials: 3 index cards with a date written in words (April 8, 1955; March 18, 1991; August 14, 2006) and 3 index cards with a date written in digits (7/4/2005, 10/11/2003, 1/24/1984)

- Give the child a card with the date written in words, and ask him or her to write the date using digits.
- Give the child a card with the date written in digits and ask him or her to write the full date using words.
- Repeat with the remaining cards.

Advanced Learners

- Extension and Enrichment problem for Guided Class Practice 47, problem 3 (Extension and Enrichment Student Problem [page 10] at www.SaxonMathResources.com)

Extension and Enrichment Problem	Answer
Kiana was born on September 16, 2006. What will be the date of Kiana's 15th birthday? Write the date using digits.	9/16/2021

- **Creating and Reading a Bar Graph**
- **Creating and Reading a Venn Diagram**

English Learners

materials: Venn diagram chart from Lesson 48

- Show the child the Venn diagram chart.
- **Venn diagram:** a drawing of two or more intersecting circles that shows what the sets have in common
- Review with the child that a set is a group with one or more attributes in common.
- Review all the parts of the Venn diagram: the title; the two circles, their labels, and what the circles show; what the area outside of the circles shows; and what the area the circles have in common shows.
- Have the child use a complete sentence to describe a *Venn diagram.*

Spanish Cognate: Venn diagram diagrama de Venn

Inclusion Students

materials: pattern blocks (hexagon, trapezoid, triangle, parallelogram, square)

- On a sheet of paper, create a 2-circle Venn diagram large enough to place pattern blocks; label the first circle *4-Sided* and label the second circle *Equal Sides.*
- In the 4-sided circle, place the trapezoid; in the overlapping section of the circles, place the square and parallelogram; in the equal sides circle, place the triangle and hexagon.
- Review with the child that the set of pattern blocks with four sides includes the trapezoid, square, and parallelogram and that the set of pattern blocks with equal sides includes the triangle, hexagon, square, and parallelogram.
- Ask the child which shapes the two sets have in common.

At-Risk Students

materials: pattern blocks (hexagon, trapezoid, triangle, parallelogram, square)

- On a sheet of paper, create a 2-circle Venn diagram large enough to place pattern blocks; label the first circle *4-Sided* and label the second circle *Equal Sides.*
- Ask the child to place the trapezoid; ask him or her where it should be placed, prompting with the questions: Does this shape have equal sides? and Does this shape have four sides?
- Repeat with the triangle and hexagon.
- Ask the child to place the square, prompting with the same questions and then asking: Since the square is both 4-sided and has equal sides, where should it be placed?
- Repeat with the parallelogram.
- Review that the intersecting part of the Venn diagram shows what the 2 groups have in common.

Advanced Learners

- Learning Palette Enrichment Card E5, Reading a Venn Diagram

- **Ordering Two-Digit Numbers**

English Learners

materials: index cards with two-digit numbers (from the lesson)

- **greatest:** more than all the others
- **least:** less than all the others
- Review with the child that there are many words that have the same meaning; we use the following words when comparing three or more amounts or numbers, and when we mean the value that is the highest: most, largest, and greatest. We use the following words when we mean the value that is the lowest (closest of the three to 1): least, smallest, and fewest.
- Give the child three of the index cards and ask him or her to arrange the cards in order from least to greatest.
- Ask the child which is the least and which is the greatest.
- Repeat with three other cards.
- Have the child use complete sentences to explain *greatest* and *least.*

Inclusion Students

materials: Math Folder (hundred number chart) and index cards with two-digit numbers (up to 12 cards from the lesson)

- Have the child compare and order the two-digit numbers; the child may refer to the hundred number chart if necessary.

At-Risk Students

- Math Center Activity 40

Objectives	Materials for Four Children	Teacher Directions
• Ordering Two-Digit Numbers	• 4 Learning Palettes • 4 Learning Palette Card 19	Ask each child to match the problems and answers on Learning Palette Card 19.

Advanced Learners

- Learning Palette Enrichment Card E6, Ordering Two-Digit Numbers

- **Addition Facts: Sums of 13 and 14**
- **Making an Organized List to Solve a Problem**

English Learners

materials: 15 color tiles

- **prove:** show to be true
- Remind the child that in class, proving something to be true can be achieved with logical reasoning or using objects, like manipulatives.
- Ask the child to prove that the sum of an even and odd number will always be odd; ask the child to explain his or her thinking using color tiles as well as in words.
- Have the child use a complete sentence to explain *prove.*

Spanish Cognate: prove probar

Inclusion Students

The child will practice addition facts for sums of 13 and 14.

materials: Teacher Fact Cards (Set B [4 + 9, 5 + 9, 9 + 4, 9 + 5], Set C [6 + 7, 7 + 6], Set H [5 + 8, 6 + 8, 7 + 7, 8 + 5, 8 + 6])

- Show the Teacher Fact Cards in random order; ask the child to say the answers.
- Review any facts the child misses.

At-Risk Students

materials: Problem-Solving Worksheet 50A and Performance Task Worksheet 50B

- Review with the child the problem-solving strategies and Problem-Solving Worksheet 50A and Performance Task Worksheet 50B.

Advanced Learners

- Extension and Enrichment problem for Problem-Solving Worksheet 50A (Extension and Enrichment Student Problem [page 11] at www.SaxonMathResources.com)

Extension and Enrichment Problem	Answer
Maya has a cup of pennies, a cup of nickels, and a cup of dimes. Show all the ways she can make 23¢ using her dimes, nickels, and pennies. How many different ways can Maya make 23¢?	2D, 3P; 1D, 2N, 3P; 1D, 1N, 8P; 1D, 13P; 4N, 3P; 3N, 8P; 2N, 13P; 1N, 18P; and 23P; 9 ways

Guide to Differentiated Instruction

- **Selecting the Appropriate Tool to Measure Capacity**
- **Reading a Recipe**
- **Measuring Ingredients for a Recipe**

English Learners

materials: liquid and dry measuring cups

- Show the different cups to the child.
- **measuring cup:** a container that holds a known amount, used in cooking and baking
- Review with the child that these cups are used to measure the ingredients used in a recipe.
- Point out the differences between the liquid and dry measuring cups.
- Note that most cooks and bakers have cups that come in fractions or multiples of a cup, such as 2-cup, half-cup, and quarter-cup containers.
- Explain that an ingredient like flour would generally be measured using a cup rather than a measuring spoon, because of the amount used in a recipe. Also the reverse, that something like salt, is generally measured with a spoon, because little is needed.
- Have the child use a complete sentence to explain *measuring cup*.

Inclusion Students

The child will measure ingredients.

materials: bag of rice (or something similar) and dry measuring cups

- Ask the child to find how many quarter cups are in a half cup and in a cup.
- Repeat, asking how many third cups are in a cup.

At-Risk Students

The child will measure ingredients.

materials: bag of rice (or something similar), dry measuring cups, and measuring spoons

- Ask the child to find and record how many quarter cups are in a half cup and in a cup, tablespoons are in a quarter cup, and teaspoons are in a tablespoon. (2, 4, 4, 3)

Advanced Learners

- Ask the child if he or she has ever used or seen somebody use a recipe.
- Have the child write the measuring tools that were used and what they were used to measure.

- **Counting Dimes, Nickels, and Pennies**

English Learners

materials: 16 color tiles, a dime, nickel, and penny

- **most:** the largest amount; greater than all the others
- Place the color tiles into groups containing 5, 10, and 1 tile; point to the middle group and explain that this group has the most tiles.
- Give the child the three coins.
- Ask him or her to match the three coins with the three groups of tiles.
- Ask him or her which coin is worth the most and to explain why. (The child may use the color tiles to help with the explanation, if he or she desires.)
- Have the child use a complete sentence to explain *most.*

Inclusion Students

materials: 5 dimes, 10 nickels, and 30 pennies in a container

- Ask the child to identify and sort the coins.
- Ask the child the value of each type of coin in cents: the value of a penny, dime, and nickel.
- Ask him or her to divide the pennies into groups of 10.
- Ask him or her to show how many nickels equals this number of pennies; ask how many dimes equals this number of pennies.
- Collect all the coins and then show the child 2 dimes, 2 nickels, and four pennies; ask him or her how much money this is. (34¢)

At-Risk Students

- Math Center Activity 42

Objectives	Materials for Four Children	Teacher Directions
• Counting Dimes, Nickels, and Pennies	• 4 Learning Palettes • 4 Learning Palette Card 20	Ask each child to match the problems and answers on Learning Palette Card 20.

Advanced Learners

materials: 2 dimes, 4 nickels, and 20 pennies in a container

- Ask the child to find and record as many combinations of coins as possible that equal 20¢. (20P; 15P, 1N; 10P, 2N; 10P, 1D; 5P, 3N; 5P, 1N, 1D; 4N; 2N, 1D; 2D)

- **Identifying a Line of Symmetry**
- **Creating a Symmetrical Design**

English Learners

materials: pattern blocks and paper shapes (heart and rectangle with a drawn line of symmetry)

- Show the child the paper shapes; point to the line of symmetry on each and explain that the two sides of the shapes are symmetrical.
- **line of symmetry:** an imaginary line that divides a figure into two identical sides
- **symmetrical:** having two identical sides that can be divided by an imaginary line
- Fold a sheet of paper in half and place several pattern blocks along one side of the fold line.
- Place matching blocks on the other side and explain that the fold line is the line of symmetry.
- Explain that to be symmetrical, the two sides should be *mirror images* of each other.
- Have the child use complete sentences to explain *line of symmetry* and *symmetrical.*

Spanish Cognate: symmetrical simétrico

Inclusion Students

The child will review a line of symmetry and finish creating a symmetrical design.

materials: pattern blocks

- Fold a sheet of 8.5" × 11" paper in half vertically.
- On the sheet, place pattern blocks on one side of the fold line, creating a colorful pattern.
- Explain to the child that the fold line will be the line of symmetry.
- Ask him or her to place matching blocks on the other side of the line to create a *mirror image* of the first side; assist the child as needed.
- When the figure is completed, review how it shows a symmetrical design.

At-Risk Students

- Math Center Activity 44

Objectives	Materials for Four Children	Teacher Directions
• Identifying Lines of Symmetry	• 4 Learning Palettes • 4 Learning Palette Card 21	Ask each child to match the problems and answers on Learning Palette Card 21.

Advanced Learners

- Extend and Challenge CD, Activity 5, Symmetry
- Have the child complete Extend and Challenge CD Activity 5 Worksheet.
- After the child finishes the activity, use the question(s) on the recording form to evaluate his or her understanding. (Monitoring Student Progress binder, Extend and Challenge CD booklet)

• Adding Two-Digit Numbers Using Dimes and Pennies (Part 1)

English Learners

- **vowel:** one of the following letters of the alphabet: a, e, i, o, u, and sometimes y
- Point to the vowels on the classroom alphabet.
- Ask the child to print his or her first and last names; then have the child underline the vowels in the names.
- Review with the child the pronunciation of *vowel* and all the letters that are vowels.
- Have the child use a complete sentence to explain *vowel*.

Inclusion Students

materials: a cup of 5 dimes and 9 pennies, and 4 index cards with the following amounts: 13¢, 15¢, 22¢, 34¢

- Ask the child to select any two cards at a time and add the two amounts using dimes and pennies.
- On a sheet of paper, create columns labeled dimes and pennies, and then write the two amounts being added, including the cent symbol.
- Review the addition of pennies and dimes to find the sum.
- Repeat with other pairs of cards.

At-Risk Students

- *Saxon Math 1,* Extend and Challenge CD, Activity 11, Let's Go Shopping
- Ask the child the following question: If you spend 1 dime and 3 pennies for one item and 2 dimes and 5 pennies for another item, how could you find out how much money you spent in all?

Advanced Learners

- Extension and Enrichment problem for Guided Class Practice 53, problem 2 (Extension and Enrichment Student Problem [page 12] at www.SaxonMathResources.com)

Extension and Enrichment Problem	Answer
Print every letter of the alphabet using capital (uppercase) letters. Do any of the letters have two lines of symmetry? Which ones?	yes; H, I, O, X

- **Adding Two-Digit Numbers Using Dimes and Pennies (Part 2)**

English Learners

- Write 13 – 8 = 5 on the board.
- **subtraction:** an operation that takes one number away from another to find the difference
- Point to the problem on the board and review that the symbol (minus sign) indicates subtraction.
- Remind the child that subtraction is the description of a some, some went away problem.
- Ask the child to make up a some, some went away story problem using the numbers 13, 8, and 5.
- Ask the child to identify the problem as an addition or subtraction problem; ask how he or she knows.

Inclusion Students

materials: a cup of 5 dimes and 9 pennies, and 4 index cards with the following amounts: 12¢, 24¢, 25¢, 33¢

- Ask the child to select any two cards at a time and add the two amounts using dimes and pennies.
- On a sheet of paper, create columns labeled dimes and pennies, and then write the two amounts being added, including the cent symbol.
- Review the addition of pennies and dimes to find the sum.
- Repeat with other pairs of cards.

At-Risk Students

materials: pennies, dimes, and on a sheet of paper make 3 dimes-and-pennies tables similar to those on Lesson Worksheet 53

- Read the following problem to the child: Bao had 78¢ in his pocket. Then he found 11¢ in his room. How much money does he have now?
- Have the child solve the problem by both adding pennies and dimes, and by filling in one of the tables.
- Repeat with the following: (1) Maria has 61¢ in her piggy bank and 22¢ in her purse; how much money does she have in all? (2) Franklin had 34¢ in his pocket and found a dime; how much money does he have now?

Advanced Learners

Journal Writing (Overview 6)

- Have the child write a some, some more story using dimes and pennies.

- **Addition Facts: Sums of 15, 16, 17, and 18**

English Learners

- On the board write the following hints (clues) to find out Pavlo's favorite number: Pavlo's favorite number is between 14 and 18. It does not end in 5. It is not an even number.
- **hint:** a tip or clue; helpful information
- Explain to the child that a hint is helpful information used to solve a problem.
- Write the numbers 14 through 18 on the board below the hints.
- Ask the child to read you the hints to solve the problem.
- Ask him or her to follow the hints and to tell you Pavlo's favorite number. (17)
- Ask the child to give you some clues so you can guess his or her age. Have him or her use the word *hint* in each sentence. (Example: One hint is ...)

Inclusion Students

materials: 4 index cards, with the following numbers: 6, 7, 8, 9

- Place the cards face down on the table, then ask the child to select 2 cards at a time and say the sum.
- Repeat several times.

At-Risk Students

materials: 10 red, 10 blue, and 10 yellow color tiles

- Ask the child to show the groups of color tiles as you read the following story: There are three baskets of flowers in a flower shop; one basket contains 8 blue flowers, one basket contains 7 red flowers, and the third basket contains 9 yellow flowers.
- Ask the child the following questions:

1. How many yellow and red flowers are there altogether? (9 + 7 = 16)
2. How many red and blue flowers are there altogether? (7 + 8 = 15)
3. How many of the fewest and most flowers are there altogether? (7 + 9 = 16)
4. If the florist added two more red flowers to the basket of red flowers, how many are in that basket now? (7 + 2 = 9)
5. If a customer asked for the new basket of red flowers and the basket of yellow flowers, how many flowers would the customer get? (9 + 9 = 18)

• Measuring to the Nearest Foot

English Learners

materials: ruler

- Give the child the ruler and have him or her examine it.
- **foot:** a U.S. Customary System unit used to measure length, height, or distance; ft
- **feet:** more than one foot
- Remind the child that a standard ruler is a foot long and that 1 foot is equal to 12 inches.
- Explain to the child that just as more than one foot on a person is called feet, more than one foot of distance is referred to as feet.
- Remind the child to begin measurement at the -0- line of the ruler, which is not always at the edge of the ruler itself.
- Help the child measure a desktop or table with the ruler.
- Have him or her use a sentence to tell you the measurement to the nearest foot.

Inclusion Students

materials: rulers and classroom items to measure

- Help the child measure several classroom items, such as a desktop, table, bookshelf, the width of a door or window, or several floor tiles.
- Ask the child to tell you his or her results to the nearest foot.

At-Risk Students

materials: rulers and classroom items to measure

- Have the child measure several classroom items, such as a desktop, table, bookshelf, the width of a door or window, or several floor tiles.
- Ask the child to record his or her results to the nearest foot; tell him or her to include the unit *ft* when writing the results.

Advanced Learners

materials: retractable tape measure

- Have the child make a chart so he or she can record the height of family members and bring the information to class.
- Ask the child to measure and record the heights of his or her family to the nearest foot; tell him or her to include the unit *ft* when writing the results.
- Have the child write about his or her discoveries (tallest, shortest, etc.).

- **Drawing a Number Line**
- **Drawing Line Segments to the Nearest Inch**
- **Locating Points on a Number Line**

English Learners

- Refer to the classroom number line as you review with the child: note that while this number line does not display arrows, the numbers continue in both directions forever.
- **line segment:** a straight path usually marked with two endpoints
- Review **number line:** a straight path marked with numbers at equal spaces and arrows at each end
- Remind the child that a line segment is part of a line, such as a number line, but that it has a beginning and an end.
- Point to the part of the number line between numbers 5 and 6; explain that this little portion of the number line is a line segment.
- Point to the part of the number line between numbers 5 and 12; explain that this also is a line segment, just a longer line segment.
- Have the child identify the number line and a line segment by pointing to them and using sentences to describe them.

Inclusion Students

materials: classroom number line and Math Folder (hundred number chart)

- Have the child locate various numbers on both the classroom number line and the hundred number chart.
- Ask him or her to explain how a number line is different from a hundred number chart.

At-Risk Students

materials: ruler and color tile

- Have the child draw a number line, labeled from 0 to 10, using the ruler and color tile to space the numbers evenly.
- On the number line, have the child place a dot at 2 and label it point A.
- Have the child place a dot at 5 and label it point B.

Advanced Learners

- Draw a number line showing negative numbers; point out that there are negative numbers on the Meeting Board thermometer and on the classroom number line.
- Review with the child that numbers on the number line continue in both directions forever.
- Explain that the distance of a negative number from 0 equals the distance of the same positive number.
  (Example: –5 is 5 units in the left-hand direction just as 5 is 5 units in the right-hand direction.)
- Ask the child how many units –2 is from 0; … how many units –4 is from 0.

• Making Polygons on a Geoboard
• Identifying the Angles of a Polygon

English Learners

materials: geoboard and geobands

- Create a triangle on the geoboard and point to each angle.
- Explain that the angles are the corners of the polygon.
- **angle:** the figure formed by two rays extending from a common endpoint
- Create a square and ask the child for the number of angles; repeat with a hexagon.
- Have the child create a polygon with three angles.
- Have the child use a complete sentence to explain *angle.*

Spanish Cognate: angle ángulo

Inclusion Students

The child will make polygons on a geoboard and identify the angles of polygons.

materials: two geoboards and geobands

- Create a square; have the child identify the figure and copy it on his or her geoboard.
- Ask the child for the number of angles in the square.
- Repeat with a triangle, parallelogram and hexagon.
- Ask the child what he or she has noticed about the number of sides and the number of angles in a polygon such as a square. (They are equal.)

At-Risk Students

- Math Center Activity 45

Objectives	Materials for Four Children	Teacher Directions
• Making Polygons on a Geoboard • Making a Congruent Shape	• 8 geoboards with geobands • 1 index card *Write on the index card: "Make a polygon with 3 angles. Make a polygon with 4 angles. Make a polygon with 5 angles."*	Pair the children. Ask each child to make a polygon with 3 angles on a geoboard. Ask each child to use another geoboard and copy his/her partner's polygon. Ask the children to repeat the activity making polygons with 4 angles and 5 angles.

Advanced Learners

- Extension and Enrichment problem for Guided Class Practice 57, problem 5 (Extension and Enrichment Student Problem [page 12] at www.SaxonMathResources.com)

Extension and Enrichment Problem	Answer
Draw a rectangle with sides of 2 inches, 4 inches, 2 inches, and 4 inches. Add all of the sides together. What is the sum? Label your answer.	a rectangle with lengths of 2", 4", 2", and 4"; 12" or 12 inches

- **Adding Three or More Single-Digit Numbers**
- **Identifying the Associative Property of Addition**

English Learners

- Write (3 + 4) + 1 = 8 on the board; beside it write 3 + (4 + 1) = 8; underneath the number sentences write: associative property of addition.
- **associative property of addition:** the property that states that changing the grouping of the addends does not affect the sum.
- Tell the child that this property of addition may be remembered using the following word clue: when something is associated with something else, it is grouped with that thing; think of the numbers being grouped together, as if they were in a club or *association.*
- Erase the label from under the equations on the board.
- Have the child use a complete sentence to explain the *associative property of addition.*

Inclusion Students

The child will use the associative property of addition.

materials: color tiles and 4 index cards with one of the following written on each: 5 lambs, 7 goats, 3 rabbits, 1 pony

- Ask the child to place the cards in a line.
- Read the following story: Tasha went to a petting zoo. She petted 5 lambs, 7 goats, 3 rabbits, and 1 pony. How many animals did Tasha pet altogether?
- Ask the child to use the cards to find the answer; color tiles may be used, if needed.
- Ask the child if it makes any difference in what order the animals are added or which two animals are added together first; place the cards in a different order and ask the child to add them again.

At-Risk Students

The child will add three or more single-digit numbers.

materials: playing cards (A–9 of clubs and hearts)

- Review that an ace is equal to 1.
- Ask the child to select 3 of the cards, place them in a column, and determine the sum.
- Then have the child place the numbers in a different order and find the sum again.
- Explain that the sum is the same regardless of the order of the addends, or which two addends are added together first, because of the commutative and associative properties of addition.
- Have the child select 3 or 4 new cards and find the sum of the cards.

Advanced Learners

- Learning Palette Enrichment Card E7, Reading a Chart

- **Naming a Fractional Part of a Set**

English Learners

materials: color tiles (blue and red)

- On the board write 2/3, 3/6, and 7/10.
- Review **fraction:** a number that tells the number of parts (numerator) of a whole (denominator)
- Point to each fraction and remind the child that the bottom number in a fraction shows how many parts make up the whole.
- Show the child 2/3 using the 1 red and 2 blue color tiles.
- Have the child repeat the activity with 3/6 and 7/10.
- Have the child explain *fraction* using a complete sentence.

Spanish Cognate: fraction fracción

Inclusion Students

materials: 6 index cards; on each of 3 index cards draw 12 small circles; on the first card shade 1/2 of the circles (6), on the second shade 1/4 of the circles (3), and on the third shade 1/6 of the circles (2); on the remaining 3 cards write one of these fractions: 1/2, 1/4, 1/6

- Have the child match the index cards showing the fractions with the cards showing the shaded parts of a set.

At-Risk Students

- Math Center Activity 46

Objectives	Materials for Four Children	Teacher Directions
• Identifying a Fractional Part of a Whole	• 4 Learning Palettes • 4 Learning Palette Card 22	Ask each child to match the problems and answers on Learning Palette Card 22.

Advanced Learners

- Learning Palette Enrichment Card E8, Comparing Fractions

Guide to Differentiated Instruction

- **Subtracting 0 Facts**
- **Drawing a Picture to Solve a Problem**
- **Looking for a Pattern to Solve a Problem**

English Learners

- Write 15 – 8 = 7 on the board.
- **minus:** a word used between numbers that means to subtract the second number from the first number
- Review with the child that the – symbol, which indicates subtraction, is call the minus sign.
- Ask the child another term for subtraction. (some, some went away)
- Ask him or her to read the equation on the board.

Inclusion Students

The child will practice subtracting 0 facts.

materials: set of Learning Wrap-Ups, #10

- Have the child say the subtracting 0 facts aloud as he or she wraps the board.

At-Risk Students

materials: Problem-Solving Worksheet 60A and Performance Task Worksheet 60B

- Review with the child the problem-solving strategies and Problem-Solving Worksheet 60A and Performance Task Worksheet 60B.

Advanced Learners

- Extension and Enrichment problem for Problem-Solving Worksheet 60A (Extension and Enrichment Student Problem [page 13] at www.SaxonMathResources.com)

Extension and Enrichment Problem	Answer
Suzanne wants to make a necklace with a blue, blue, red, green, green, yellow repeating pattern. She will use 18 beads to make her necklace. Show how many of each color bead she will need.	a picture showing the repeating pattern of beads: BBRGGY BBRGGY BBRGGY (6 blue, 3 red, 6 green, 3 yellow)

• Identifying and Creating Congruent Shapes

English Learners

materials: 2 geoboards and geobands

- Create two congruent triangles, one on each of the geoboards; and explain that since they have the same size and shape, they are congruent.
- **congruent:** having the same size and shape
- Rotate one of the geoboards, and explain that the triangles are still congruent even when one is turned a different way than the other.
- Create another triangle shape and have the child copy the shape on the second geoboard; say that since the child's copy is the same size and shape, the triangles are congruent.
- Repeat with a rectangle.
- Have the child use a complete sentence to explain *congruent.*

Spanish Cognate: congruent congruente

Inclusion Students

materials: 2 geoboards and geobands

- Create a square shape and have the child copy the shape on his or her geoboard.
- Explain that because the squares are the same size and shape, they are congruent.
- Repeat with a rectangle and triangle.

At-Risk Students

- Math Center Activity 49

Objectives	Materials for Four Children	Teacher Directions
• Identifying Congruent Shapes	• 4 Learning Palettes • 4 Learning Palette Card 23	Ask each child to match the problems and answers on Learning Palette Card 23.

Advanced Learners

materials: 2 geoboards and geobands

- Create a trapezoid using a geoboard and geobands.
- Have the child create a congruent trapezoid using the second geoboard.
- Rotate one of the geoboards and explain that the trapezoids are still congruent even when one is turned a different way than the other.
- Repeat the activity with a parallelogram and a hexagon.

- **Adding Two-Digit Numbers With Regrouping, Using Dimes and Pennies (Part 1)**

English Learners

- Write the following addition problem vertically on the board: 16¢ + 14¢ = 30¢.
- **regrouping:** creating different groups or sets from those that already exist
- Review that when the number of pennies in a group is 10 or more, 10 of the pennies may be traded for one dime.
- Point to the example and review regrouping; show that when the 6 cents and 4 cents are added to make 10 cents, that a dime is added to the tens' column or is grouped with the tens.
- If necessary, show more examples.
- Ask the child to explain using a complete sentence what it is called when you create different groups from groups that are already there.
- Have the child give an example.

Spanish Cognate: regrouping reagrupación

Inclusion Students

materials: Workmat 28, a cup of 13 pennies, and a cup of 4 dimes

- Write the following addition problem vertically on the board: 16¢ + 14¢.
- Ask the child to show 16¢ on the workmat using dimes and pennies.
- Ask him or her to then add 14¢ to the workmat using dimes and pennies.
- Ask the child if it is possible to trade the pennies for a dime.
- Have him or her do this as you show this on the written problem.
- Show the sum (30¢) as the child counts the three dimes.
- Repeat these steps with 28¢ + 15¢.

At-Risk Students

materials: Workmat 28, a cup of 13 pennies, and a cup of 5 dimes

- Write the following addition problem vertically on the board: 16¢ + 14¢.
- Ask the child to show 16¢ on the workmat using dimes and pennies.
- Ask him or her to then add 14¢ to the workmat using dimes and pennies.
- Ask the child if it is possible to trade the pennies for a dime.
- Have him or her do this as you show this on the written problem.
- Show the sum (30¢) as the child counts the three dimes.
- Repeat these steps with 18¢ + 22¢, 33¢ + 19¢, and 28¢ + 15¢.

- **Adding Two-Digit Numbers With Regrouping, Using Dimes and Pennies (Part 2)**

English Learners

materials: a picture of a building with columns

- **column:** a vertical arrangement of items
- Show the child the picture of the building with the columns, pointing out that the columns are vertical (go up and down).
- Write a vertical addition problem with two-digit addends; point out the tens and ones columns.
- Remind the child that it is important to write the numbers in straight columns, so that it is clear in which column the numbers belong.
- Write an addition problem in crooked columns to show how confusing it is if the columns of numbers are not straight.
- Ask the child to write the addition problem 15 + 10 vertically; ask him or her to explain what the two vertical lines of numbers are and to find the sum. (ones' or pennies column, tens' or dimes column, 25)

Spanish Cognate: column columna

Inclusion Students

materials: Workmat 28, a cup of 20 pennies, a cup of 5 dimes, and 3 classroom items with price tags (made from self-stick notes) with the following prices: 6¢, 19¢, 28¢

- Ask the child to select any two of the items and find the total cost.
- The child should show the number of pennies and dimes needed for the first item on his or her workmat; he or she should then show the number of pennies and dimes needed for the second item.
- The child should trade ten pennies for a dime, and then count the coins to find the total cost of the two items.
- Repeat with another pair of items.

At-Risk Students

materials: Workmat 28, a cup of 20 pennies, a cup of 8 dimes, and 5 classroom items with price tags (made from self-stick notes) with the following prices: 6¢, 15¢, 19¢, 28¢, 37¢

- Ask the child to select any two of the items and find the total cost.
- The child should show the number of pennies and dimes needed for the first item on his or her workmat; he or she should then show the number of pennies and dimes needed for the second item.
- The child should trade ten pennies for a dime, and then count the coins to find the total cost of the two items.
- Repeat with other pairs of items.

Advanced Learners

- Extend and Challenge CD, Activity 6, Let's Go Shopping
- After the child finishes the activity, use the question(s) on the recording form to evaluate his or her understanding. (Monitoring Student Progress binder, Extend and Challenge CD booklet)

• Using the Addition Algorithm (Part 1)

English Learners

materials: 13 pennies and 1 dime

- **extra:** more than needed for the purpose; leftover or additional
- Show the child how to trade ten pennies for a dime; explain that while starting with 13 pennies, there are only 3 pennies left after the trade; these pennies are *extra.*
- Give the child 12 pennies and ask him or her to trade the necessary number for a dime.
- Ask him or her to explain using a complete sentence what the additional pennies are called and how many there are.

Spanish Cognate: extra extra

Inclusion Students

materials: Workmat 28, a cup of 20 pennies, a cup of 5 dimes, and 3 classroom items with price tags (made from self-stick notes) with the following prices: 7¢, 15¢, 26¢

- Ask the child to select any two of the items and find the total cost.
- The child should show the number of pennies and dimes needed for the first item on his or her workmat; he or she should then repeat for the second item.
- The child should trade ten pennies for a dime, and then count the coins to find the total cost of the two items.
- Repeat with another pair of items.

At-Risk Students

- Draw an addition chart (like that used in Lesson 63) on the board and review the chart with the child, emphasizing the importance of keeping separate columns for the pennies and dimes.
- Ask the child to copy the chart onto a piece of paper.
- Write 26¢ + 44¢ = on the board; ask the child to write this addition problem on the chart, being sure to separate the pennies and dimes.
- Ask him or her to complete the addition, first adding the pennies, showing the trade of pennies for a dime, and then adding the dimes.
- Repeat with 35¢ + 28¢ = .

Advanced Learners

- Extension and Enrichment problem for Guided Class Practice 63, problem 3 (Extension and Enrichment Student Problem [page 14] at www.SaxonMathResources.com)

Extension and Enrichment Problem	Answer
Draw or trace a triangle, a trapezoid, a hexagon, and a parallelogram. Which has the greatest number of angles? Compare the number of angles to the number of sides of each shape.	drawings of a triangle, a trapezoid, a hexagon, and a parallelogram; hexagon; the number of angles is the same as the number of sides

- **Using the Addition Algorithm (Part 2)**

English Learners

materials: on six separate small pieces of paper, draw the following figures: 2 triangles, 2 squares, and 2 circles; each pair of figures should look alike

- **match:** the same or equal in appearance or value
- Place the drawn figures in front of the child; point to the two circle drawings as you explain that these figures match each other because they are alike.
- Ask the child to match the remaining pairs of figures.
- Explain that during a lesson, the children in the class are sometimes asked if what they have written on their worksheets matches what is written on the board; this means they are being asked if the problem on their worksheet looks just the same.
- Ask the child to pair the figures on the pieces of paper again and to use a complete sentence to explain why he or she chose the pairs as he or she did. (they match)

Inclusion Students

materials: 2 cups of coins, one with 12¢ and one with 26¢ (using dimes and pennies); index cards with the following amounts written: 12¢, 15¢, 21¢, 26¢, 38¢, 41¢

- Give the child the first cup of coins; ask him or her to count the money and find the card that equals that amount.
- Repeat with the second cup of coins.
- Ask the child to add the two cups of coins together and find the total.
- Ask him or her to select the card that shows this amount.

At-Risk Students

- Math Center Activity 51

Objectives	Materials for Four Children	Teacher Directions
• Adding Two-Digit Numbers	• 4 Learning Palettes • 4 Learning Palette Card 24	Ask each child to match the problems and answers on Learning Palette Card 24.

Advanced Learners

- Write 35¢ + 27¢ + 15¢ = on the board.
- Ask the child to write a short some, some more story with these numbers and to find the sum.

• Subtracting 1 Facts

English Learners

- Write 6 – 1 = 5 on the board.
- **subtract:** to count backward; take away
- Explain that subtract is an action word; it is the act of taking away one number from another.
- Review that the word *minus* is used in subtraction problems when the dash, or minus sign, is shown in a number sentence.
- Ask the child to read the problem on the board.
- Ask the child to explain what action is taking place, and what the action is called.

Inclusion Students

materials: Teacher Fact Cards (Set I, Subtracting 1 Facts) and set of Learning Wrap-Ups, #1

- Show the Teacher Fact Cards in random order; ask the child to say the answers.
- Have the child say the subtracting 1 facts aloud as he or she wraps the board.

At-Risk Students

- Math Center Activity 52

Objectives	Materials for Four Children	Teacher Directions
• Subtraction Facts: Subtracting 1	4 sets of Wrap-Ups, #1	Allow the children to use Wrap-Up 1 to practice subtracting 1 from a number.

Advanced Learners

- Have the child write a number sentence for the following story: Mrs. Mills bought twelve chocolate chip cookies. Maria ate four of the cookies. How many cookies were left?
- Write 16 – 8 = 8. Have the child write a short some, some went away story about this number sentence.

- **Identifying and Creating Similar Shapes and Designs**

English Learners

materials: 2 geoboards and geobands; make a large square on one board and a small square on the other

- **similar:** things or designs that are alike in shape
- Show the child the large and small squares on the geoboards; explain that the squares are similar because their shape is the same, even though their size is not.
- Stress that to be similar, things do not have to be the same size, but must be the same shape.
- On one geoboard, change the square to a rectangle; ask the child if the shapes are still similar.
- Have the child use a complete sentence to explain *similar.*

Spanish Cognate: similar similar

Inclusion Students

materials: 2 geoboards and geobands

- Create a triangle on one geoboard and have the child copy the shape on his or her geoboard.
- Explain that when shapes are identical in size and shape, they are congruent; rotate one of the geoboards and point out that the triangles are still congruent, even though one of the figures is turned.
- Repeat with a rectangle.
- Then on one geoboard leave the original rectangle, but on the other make a similar rectangle that is either larger or smaller; point out that now the rectangles are similar, but no longer congruent, since their sizes are different.
- Repeat with a square.

At-Risk Students

- Math Center Activity 53

Objectives	Materials for Four Children	Teacher Directions
• Identifying and Creating Similar Shapes and Designs	• 4 geoboards with geobands • Center 53 Master *Make 1 copy for each child in class.*	Pair the children. Ask each child to make the first shape on Center 53 Master on a geoboard. Ask partners to check each other's geoboards. Ask the children to repeat the activity for the next two shapes. Then ask each child to make his/her own design on a geoboard and record it on the fourth geoboard on Center 53 Master. Ask each child to try to make his/her partner's design on a geoboard.

• Creating and Reading a Venn Diagram

English Learners

materials: Chart 66 (Vegetables We Like)

- Review **Venn diagram:** a drawing of two or more intersecting circles that shows what the sets have in common
- Review with the child that a set is a group with one or more common attributes; remind him or her that when sorting objects into groups he or she created sets.
- Review all the parts of the Venn diagram: the title; the two circles, their labels, and what the circles show; what the area outside of the circles shows; and what the area the circles have in common shows.
- Ask the child to tell you a fact about the Vegetables We Like chart.
- Ask the child to use a complete sentence to tell you what type of chart it is.

Spanish Cognate: Venn diagram diagrama de Venn

Inclusion Students

materials: tangram pieces: medium and small blue triangles (1 each), blue and green squares, blue and green parallelograms

- On a sheet of paper create a 2-circle Venn diagram large enough to place the tangram pieces; label the left side *Blue* and the right side *4-Sided.*
- As you discuss the reason for the placement of the pieces with the child, put the medium blue triangle in the left circle, the green square in the right circle, and the blue parallelogram in the overlapping section.
- Ask the child to place the remaining three tangram pieces and ask him or her to descibe why the pieces were placed as they were. Assist the child, if necessary, and talk about incorrectly placed pieces. (small blue triangle, left circle; green parallelogram, right circle; blue square, overlapping section)

At-Risk Students

- Math Center Activity 54

Objectives	Materials for Four Children	Teacher Directions
• Reading a Venn Diagram	• 4 Learning Palettes • 4 Learning Palette Card 25	Ask each child to match the problems and answers on Learning Palette Card 25.

Advanced Learners

- Learning Palette Enrichment Card E9, Reading a Venn Diagram

Guide to Differentiated Instruction

- **Identifying a.m. and p.m.**
- **Identifying Noon and Midnight**
- **Identifying Dozen and Half Dozen**

English Learners

materials: egg carton, 6 yellow and 6 green color tiles

- **dozen:** a set of twelve
- **half dozen:** a set of six
- Show the egg carton to the child and show that there are places for 12, or a dozen, eggs. If the word *dozen* is printed on the carton, show that to the child.
- Put the color tiles in the carton, one color on the left side and the other on the right. Show that half of 12 tiles, or 6 tiles, equals a half dozen.
- Ask the child how many sets of a half dozen can be made from one dozen; ask him or her to use the color tiles to help with the explanation, if necessary.
- Ask the child to use a complete sentence to explain what a group of 12 items is called. Ask which he or she would like more: a dozen roses or a dozen cookies; ask why.

Spanish Cognate: dozen docena

Inclusion Students

The child will identify dozen and half dozen.

materials: 14 red color tiles

- Remind the child that a dozen means a set of twelve.
- Ask the child to show you a dozen red color tiles.
- Ask him or her to show a half dozen tiles from that set.
- Ask the child to tell you how many tiles make a half dozen; ask how many half dozens make a dozen.

At-Risk Students

The child will identify dozen and half dozen.

- Ask the child to draw a set of one dozen items, such as doughnuts, eggs, or flowers.
- Ask him or her to color half of the set and to write the number of how many he or she colored (6); ask what the set of six is called. (half dozen)

Advanced Learners

- Extension and Enrichment problem for Guided Class Practice 67, problem 3 (Extension and Enrichment Student Problem [page 15] at www.SaxonMathResources.com)

Extension and Enrichment Problem	Answer
Tanner woke up at 7:30 in the morning. He stayed awake for 14 hours before going to bed. What time did Tanner go to bed? Was it morning or night when he went to bed? If Tanner sleeps 9 hours, what time will he wake up in the morning?	9:30 p.m.; night; 6:30 a.m.

• Adding Three Two-Digit Numbers

English Learners

materials: sales slip (receipt) from one or more stores

- **sales slip:** a printed statement that shows something has been paid for; receipt
- Show the sales slip(s) to the child as you explain that they are printed at the store whenever something is bought.
- Point out the things usually shown on a sales slip: the name of the store, the date, the list of items that were purchased and their individual prices, and the total cost of all the items paid for.
- Have the child explain to you, as if you didn't know, what a *sales slip* is.

Inclusion Students

materials: classroom items with price tags (from Lesson 68), a cup of 10 dimes, a cup of 30 pennies, and a copy of Center 55 Master

- Ask the child to choose two classroom items and write their cost on the Center 55 Master.
- Ask the child to find the total cost and check the answer using pennies and dimes.
- Repeat this activity with two different items. Make sure at least one pair of items requires regrouping.

At-Risk Students

- Math Center Activity 55

Objectives	Materials for Four Children	Teacher Directions
• Adding Two-Digit Numbers With Regrouping Using Dimes and Pennies	• classroom items with price tags (from Lesson 68) • 2 cups of 10 dimes • 2 cups of 30 pennies • Center 55 Master *Make 1 copy for each child in class.*	Ask each child to choose three classroom items and record his/her cost on Center 55 Master. Ask each child to find the total cost and check the answer using dimes and pennies. Repeat the activity two more times using different items.

Advanced Learners

- Extension and Enrichment problem for Guided Class Practice 68, problem 4 (Extension and Enrichment Student Problem [page 15] at www.SaxonMathResources.com)

Extension and Enrichment Problem	Answer
At the school store, Sondra bought a ruler, an eraser, 4 blue pens, 8 orange pencils, and a half-dozen green pencils. Rajat gave her 3 red pencils and Julianne gave her 2 more green pencils. Draw a picture to show how many green pencils she has now. Draw tally marks to show the total number of pencils Sondra has now.	a picture showing 8 green pencils; 19 tally marks

- **Reading a Thermometer to the Nearest 2 Degrees Fahrenheit**

English Learners

- On the board write 32°F.
- **Fahrenheit:** a scale for measuring temperature named for Gabriel Fahrenheit; on the Fahrenheit scale, the freezing point of water is 32° and the boiling point of water is 212°
- Review that temperature may be read and recorded on different scales, but in the United States, the temperature scale that is used most often in daily life is the Fahrenheit scale; explain that weather reports in this country will give the temperature in degrees Fahrenheit.
- Explain to the child that it is important when writing the temperature to show a small, raised circle to indicate degrees and to show the scale in which the temperature was measured; a capital F is used to show Fahrenheit.
- Point to the temperature on the board and ask the child to read it, including the scale; ask the child if he or she recognizes that temperature as important; if not, explain that 32°F is the freezing point of water, the temperature at which water turns to ice.

Spanish Cognate: Fahrenheit Fahrenheit

Inclusion Students

materials: the Meeting Board thermometer

- Review that on the Meeting Board thermometer, each small line indicates 2 degrees.
- On the thermometer, draw 48° and ask the child to identify the temperature.
- Repeat for: −2°, 16°, 84°.
- Ask the child to draw 52° on the thermometer.
- Repeat for: 44°, 32°.
- Ask the child if he or she remembers what is special about the temperature 32°F. (freezing point of water)

At-Risk Students

- Math Center Activity 56.

Objectives	Materials for Four Children	Teacher Directions
• Reading a Thermometer to the Nearest Five Degrees	• 4 Learning Palettes • 4 Learning Palette Card 26	Ask each child to match the problems and answers on Learning Palette Card 26.

Advanced Learners

Journal Writing (Overview 7)

- Have the child write a paragraph about what he or she would wear if it was 30°F and where he or she would go; have the child write a second paragraph about what he or she would wear if it was 80°F and where he or she would go.

Guide to Differentiated Instruction

- **Subtracting 2 Facts**
- **Solving a Problem by Guessing and Checking**

English Learners

materials: put several color tiles in a small zip-top bag

- **guess:** to make an estimate based on given information
- Say to the child that you would like to know how many tiles are in the bag, but don't want to count them.
- Explain that you can roughly estimate, or guess, the number of tiles based on the size of the tiles and the size of the bag.
- Ask the child to guess the number of tiles in the bag.
- Ask the child to remove and count the tiles; ask how accurate the guess was. (If the guess was very far off, explain that guessing is a skill that gets better with practice.)
- Ask the child to use a complete sentence to explain what word is used when you want to estimate something.

Inclusion Students

The child will practice subtracting 2 facts.

materials: Teacher Fact Cards (Set J, Subtracting 2 Facts) and set of Learning Wrap-Ups, #2

- Show the Teacher Fact Cards in random order; ask the child to say the answers.
- Have the child say the subtracting 2 facts aloud as he or she wraps the board.

At-Risk Students

materials: Problem-Solving Worksheet 70A and Performance Task Worksheet 70B

- Review with the child the problem-solving strategies and Problem-Solving Worksheet 70A and Performance Task Worksheet 70B.

Advanced Learners

- Extension and Enrichment problem for Problem-Solving Worksheet 70A (Extension and Enrichment Student Problem [page 16] at www.SaxonMathResources.com)

Extension and Enrichment Problem	Answer
At the Village School store children can buy fruit drops, mints, and licorice. Fruit drops cost 1¢ each, licorice pieces cost 2¢ each, and mints cost 3¢ each. Todd bought at least one of each. He spent 11¢ altogether. Show all the possible combinations of fruit drops, licorice, and mints he could have bought. Explain how you got your answer.	5 combinations: 2M, 2L, 1F; 2M, 1L, 3F; 1M, 3L, 2F; 1M, 2L, 4F; 1M, 1L, 6F; explanations may vary

• Identifying and Creating Overlapping Geometric Shapes

English Learners

materials: any 2 Attribute Shape pieces, geoboard, and geobands

- **overlap:** extend over to partly cover
- Place the Attribute Shapes so that one overlaps the other.
- Explain that overlap means one thing partially, but not completely, covers another; when the items cannot be seen through, like the Attribute Shapes, part of the bottom piece will still be visible.
- On the geoboard, show two rectangles that overlap; point to the area the rectangles have in common; explain that this is the area where they overlap.
- Ask the child to use a complete sentence to explain *overlap*.

Inclusion Students

materials: 2 geoboards and geobands

- For each design described, create the design on one geoboard and ask the child to copy it.
- Make a 2 × 2 square with the top extending from the top left peg to the third peg; make a second 2 × 2 square with the top starting one to the right of the top left peg and extending to the fourth peg; ask the child the shape of the entire figure and the shape of the overlapping portion. (rectangle, rectangle)
- Make a 2 × 2 square with the top extending from the top left peg to the third peg; make a second 2 × 2 square starting one diagonally down and to the right of the top left peg; ask the child the shape of the overlapping portion. (square)

At-Risk Students

materials: 2 geoboards and geobands

- For each design described, create the design on one geoboard and ask the child to copy it.
- Make a 2 × 2 square with the top extending from the top left peg to the third peg; make a second 2 × 2 square with the top starting one to the right of the top left peg and extending to the fourth peg; ask the child the shape of the entire figure and the shape of the overlapping portion. (rectangle, rectangle)
- Make a 2 × 2 square with the top extending from the top left peg to the third peg; make a second 2 × 2 square starting one diagonally down and to the right of the top left peg; ask the child the shape of the overlapping portion and whether it is congruent or similar to the two larger squares. (square, similar)
- Make a triangle with the top point on the peg one diagonally down and to the right of the top left peg and with the base along the first three pegs of the bottom row; make a mirror image of this triangle with the base along the first three pegs of the top row and the point on the peg one diagonally up and to the right of the bottom left peg; ask the child the shape of the overlapping portion. (parallelogram [rhombus])

• Using Mental Computation to Subtract 10 From a Two-Digit Number

English Learners

- Write the number sentence 47 – 10 = on the board.
- Explain to the child that when we solve a number sentence for the answer or work any mathematical problem we are using computation.
- **computation:** the act of finding a mathematical answer
- Tell the child that mathematical activities such as addition and subtraction are computations.
- Remind the child that computations may be done with the help of aids such as coins or the hundred number chart, or may be done in our heads, using mental computation.
- Ask the child why it is helpful to be able to do a computation in his or her head, without the use of aids.
- Have the child explain *computation* by completing this sentence: Mathematical computation is ....

Spanish Cognate: computation computación

Inclusion Students

materials: Math Folder (hundred number chart)

- Write 78 – 10 = on the board and ask the child to point to the number 78 on the hundred number chart.
- Ask the child to point to and tell you what number is ten less than 78; he or she may count backward by ones, if necessary.
- If his or her answer was correct, write 68 to complete the number sentence on the board.
- Ask the child where 68 is from 78 on the hundred number chart. (just above)
- Remind the child that when 10 is subtracted from a number, the ten's digit is one less.
- Repeat with the following problems: 90 – 10 = , 45 – 10 = , 21 – 10 = , 39 – 10 = .

At-Risk Students

- Math Center Activity 58

Objectives	Materials for Four Children	Teacher Directions
• Subtracting 10 From a Number Using Mental Computation	• 4 Learning Palettes • 4 Learning Palette Card 27	Ask each child to match the problems and answers on Learning Palette Card 27.

Advanced Learners

Journal Writing (Overview 8)

- Have the child write a some, some went away story problem in which he or she subtracts 10.

• Measuring and Drawing Line Segments to the Nearest Half Inch

English Learners

materials: ruler and Math Folder (inch ruler)

- **half inch:** one half of an inch
- Review with the child that when something is divided evenly into two equal parts, each part is a half.
- Show the child the ruler printed on the Math Folder; point out the line half-way between each numbered inch.
- Review that each of the two equal portions has the length of half of an inch.
- With a pencil point, point to the following marks on the printed ruler and ask the child to name them: 5", 3", 2½", 6½" and ½".
- Show the child that on the plastic ruler there are more lines between the numbered inches; show him or her that the inch lines are the longest, and that the half-inch lines (midway between) are the next longest.
- Ask the child to measure the folded Math Folder across the top using the plastic ruler and tell you the measurement. (9½")

Inclusion Students

materials: Math Folder (inch ruler) and straws cut to the following lengths: 4", 4½", 5", 5½"

- Place the straws in random order in front of the child; ask him or her to arrange the straws from shortest to longest.
- Ask the child to measure each straw and say the measurement to the half inch.

At-Risk Students

materials: ruler and paper strips cut to the following lengths: 1½", 4", 4½", 5", 7½", 8", 9½"

- Place the paper strips in random order in front of the child; ask him or her to arrange the strips from shortest to longest.
- Ask the child to measure each strip and record the measurement to the half inch.

Advanced Learners

- Extension and Enrichment problem for Guided Class Practice 72, problem 1 (Extension and Enrichment Student Problem [page 16] at www.SaxonMathResources.com)

Extension and Enrichment Problem	Answer
Simon has 4 dimes and 5 pennies. Edgar has 8 dimes and 8 pennies. Write a number sentence to show how many more dimes Edgar has than Simon. Write a number sentence to show how many more pennies Edgar has than Simon. How much more money does Edgar have than Simon? Write a number sentence to show this.	8 dimes – 4 dimes = 4 dimes; 8 pennies – 5 pennies = 3 pennies; 88¢ – 45¢ = 43¢

• Adding Two-Digit Numbers With a Sum Greater Than 100

English Learners

- Write the following numbers on the board: 67, 235, 19, 126, 488, 5, 115, 100.
- Explain to the child that if we compare the first two numbers, we know that 235 is greater than 67 because it is more than.
- **greater than:** more than
- Point to each of the numbers on the board and ask the child if the number is greater than 100.
- Ask the child to explain to you how he or she decides if one number is greater than another number.

Inclusion Students

- Write the following problems on the board vertically: 79 + 31, 85 + 68.
- For the first problem, ask the child to add the ones' column first; ask if it is necessary to trade 10 ones for 1 ten.
- Ask the child to add the tens' column (including a carried 1).
- Ask him or her whether the sum is greater than 100.
- Repeat with the other problem.

At-Risk Students

- Math Center Activity 59

Objectives	Materials for Four Children	Teacher Directions
• Adding Two-Digit Numbers	• 4 Learning Palettes • 4 Learning Palette Card 28	Ask each child to match the problems and answers on Learning Palette Card 28.

Advanced Learners

- Extension and Enrichment problem for Guided Class Practice 73, problem 1 (Extension and Enrichment Student Problem [page 17] at www.SaxonMathResources.com)

Extension and Enrichment Problem	Answer
Eleven children from Grade 1 and 16 children from Grade 2 went on a field trip to the aquarium. Four children went home early. How many children finished the field trip?	23 children

- **Representing Numbers Using Base Ten Blocks**
- **Ordering Numbers Using Base Ten Blocks**

English Learners

materials: a die, a base ten unit cube, the 1000-cube, and the set of geometric solids

- Show the child the cube from the geometric solids set and point out the six square faces.
- **cube:** a geometric solid with six congruent square faces
- Show the child the two base ten cubes—point out that size does not matter; an object is still a cube as long as the object has six square sides.
- Place all the solids in front of the child and ask him or her to select the cubes.
- Have the child use a complete sentence to describe a *cube.*

Spanish Cognate: cube cubo

Inclusion Students

The child will represent and order numbers using base ten blocks.

materials: base ten blocks and 3 index cards; write a different three-digit number on each index card

- Ask the child to place the number cards in order from least to greatest.
- Ask the child to show the value of each number with the base ten blocks.
- Ask him or her to tell you the number of hundreds, tens, and ones in each number, and to say the number.

At-Risk Students

- Math Center Activity 60

Objectives	Materials for Four Children	Teacher Directions
• Ordering Three-Digit Numbers	• 20 index cards • 4 small zip-top bags *Write a different three-digit number on each of the 20 index cards. Put a set of 5 cards in each of the zip-top bags.*	Pair the children. Ask each child to select a bag of cards and place the number cards in order from least to greatest. Have the partners check each other's card order. Then have the children combine their sets of cards and work together to order the 10 cards from least to greatest. When they finish, ask them to put five cards in each bag and repeat the activity with the other two bags of cards.

Advanced Learners

- Learning Palette Enrichment Card E10, Identifying Numbers Before, After, and Between

- **Subtracting 3 Facts**

English Learners

- Write the following subtraction problem on the board both horizontally and vertically: $8 - 3 = 5$
- **difference:** the answer to a subtraction problem
- Explain that the difference, 5, is the answer in this subtraction problem, and it shows by how much the two numbers, 8 and 3, are different.
- Write the following four number sentences on the board: $4 + 3 = 7$, $7 - 3 = 4$, $7 - 4 = 3$, $3 + 4 = 7$.
- Remind the child that the numbers 3, 4, and 7 form a fact family and from these numbers the four facts on the board can be written.
- Ask the child to point out the answer in each number sentence and identify it as either a sum or a difference.

Spanish Cognate: difference diferencia

Inclusion Students

materials: Teacher Fact Cards (Set K, Subtracting 3 Facts) and set of Learning Wrap-Ups, #3

- Show the Teacher Fact Cards in random order; ask the child to say the answers.
- Have the child say the subtracting 3 facts aloud as he or she wraps the board.

At-Risk Students

- Math Center Activity 61

Objectives	Materials for Four Children	Teacher Directions
• Subtraction Facts: Subtracting 3	• 4 sets of Wrap-Ups, #3	Allow the children to use Wrap-Up 3 to practice subtracting 3 from a number.

Guide to Differentiated Instruction

- **Identifying Gallon, Half-Gallon, Quart, and Liter Containers**
- **Estimating and Finding the Capacity of Containers**

English Learners

materials: gallon and half-gallon containers (two of each, one empty and one full of water)

- **capacity:** the amount that a container can hold
- **full:** completely filled
- Explain that a container is full when it cannot hold any more, and that when it is full, that is a measure of the capacity of the container.
- Point out that the capacity of a container (such as gallon and half-gallon) depends on its size.
- Have the child use complete sentences to explain *full* and *capacity.*

Spanish Cognate: capacity capacidad

Inclusion Students

The child will identify gallon, half-gallon, quart, liter, and cup containers.

materials: measuring cup, quart, liter, half-gallon, and gallon containers

- Remind the child that the capacity of a container depends on its size; ask him or her to put the containers in order from the one that would hold the least to the one that would hold the most.
- Review the terms cup, liter, quart, half-gallon, and gallon as you point to each container.

At-Risk Students

- Math Center Activity 62

Objectives	Materials for Four Children	Teacher Directions
• Estimating, Measuring, and Comparing the Capacity of Containers	• 1 small plastic measuring cup or other container • 5 larger plastic containers of various sizes labeled A, B, C, D, and E • rice or beans • funnel • Optional: large rimmed container • Center 61 Master *Make 1 copy for each child in class.*	Pair the children. Ask each pair to estimate and then to count the number of small cups needed to fill the larger plastic containers. Have children record the answers on Center 61 Master. Ask each pair to put the containers in order from least to greatest capacity.

Advanced Learners

Journal Writing (Overview 8)

- Ask the child to list two drinks that come in gallon containers. Have the child write which is his or her favorite and when he or she drinks it.

- **Identifying the Place Value of a Digit in a Three-Digit Number**
- **Writing a Three-Digit Number for a Model**
- **Representing Three-Digit Numbers Pictorially**

English Learners

materials: the 1000-cube, one 100-flat, one 10-stick, and ten unit cubes

- **unit:** one thing; one of a group of things
- Review with the child that the unit cube shows the value of one; remind him or her that 10 unit cubes can be traded for one 10-stick just as ten pennies can be traded for one dime.
- Write the following sentence on the board: There are ____ unit cubes in a 10-stick.
- Ask the child to read the sentence, filling in the blank with the correct value.
- Erase 10-stick and write 100-flat; ask the child to read the sentence, filling in the blank with the correct value.
- Erase 100-flat and write 1000-cube; ask the child to read the sentence, filling in the blank with the correct value.

Spanish Cognate: unit unidad

Inclusion Students

The child will identify three-digit numbers shown pictorially.

materials: 10 index cards; on five separate cards draw pictures to show the following values: 221, 315, 406, 583, 862 (using a large square for 100, a tower for 10, and a small square for 1); write the numbers as digits on the remaining cards

- Have the child match the pictures to the numbers.
- Remove the picture cards and ask the child to order the number cards from largest to smallest.
- Ask him or her to say each number.

At-Risk Students

- Math Center Activity 63

Objectives	Materials for Four Children	Teacher Directions
• Identifying a Two- or Three-Digit Number Represented by a Picture	• 4 Learning Palettes • 4 Learning Palette Card 29	Ask each child to match the problems and answers on Learning Palette Card 29.

Guide to Differentiated Instruction

- **Writing a Three-Digit Number for a Model or Picture**
- **Ordering Three-Digit Numbers**
- **Identifying the Median of a Set of Numbers**

English Learners

- Write the following numbers on the board: 317, 402, 93, 115, 268.
- **median:** the middle number in a list of numbers arranged in order from least to greatest
- Remind the child that when you want to find the median of a list of numbers, the numbers must first be placed in order.
- Ask the child to rewrite the numbers on the board in order from least to greatest.
- Ask him or her to say which number is the median; if necessary, remind him or her that *median* means the number in the middle.
- Have the child use a complete sentence to explain *median.*

Spanish Cognate: median mediana

Inclusion Students

The child will order three-digit numbers.

materials: base ten blocks, 6 index cards, and 2 small zip-top bags; write a different three-digit number on each of the 6 index cards; put a set of 3 cards in each of the zip-top bags

- Ask the child to select a bag of cards and use the base ten blocks to show each number.
- Ask him or her to order the number cards from least to greatest.
- Repeat the activity with the other bag of cards.

At-Risk Students

- Math Center Activity 65

Objectives	Materials for Four Children	Teacher Directions
• Identifying a Two- or Three-Digit Number Represented by a Picture	• 4 Learning Palettes • 4 Learning Palette Card 30	Ask each child to match the problems and answers on Learning Palette Card 30.

Advanced Learners

- Learning Palette Enrichment Card E11, Matching a Number to a Picture

• Telling and Showing Time to Five-Minute Intervals

English Learners

materials: demonstration clock

- **minute:** a unit used to measure time; *60 minutes = 1 hour*
- Review that the short hand on a clock is the hour hand and the long hand is the minute hand; remind the child that the minute hand travels one time around the face of the clock in one hour.
- Review that the dots on the demonstration clock indicate minutes; there are 60 minutes around the clockface to show the 60 minutes in one hour.
- Set the demonstration clock to 10:05 and say that the time shown may be said *ten-oh-five*; ask the child if he or she knows another way to say the time. (5 minutes past 10 or 5 minutes after 10)

Spanish Cognate: minute minuto

Inclusion Students

materials: demonstration clock; 4 student clocks showing the following times: 3:40, 5:55, 6:10, and 9:20; and 4 index cards showing these times digitally

- Ask the child to match each clock to the digital time; ask him or her to say the time.
- On the demonstration clock, show 7:05; ask the child to tell you the time shown.
- Repeat with: 4:50, 7:35.
- Write 11:10 on the board; ask the child to show this time on a student clock.
- Repeat with: 2:20, 8:15.

At-Risk Students

- Math Center Activity 66

Objectives	Materials for Four Children	Teacher Directions
• Telling Time to Five-Minute Intervals	• 4 Learning Palettes • 4 Learning Palette Card 31	Ask each child to match the problems and answers on Learning Palette Card 31.

Advanced Learners

- Extension and Enrichment problem for Guided Class Practice 78, problem 1 (Extension and Enrichment Student Problem [page 18] at www.SaxonMathResources.com)

Extension and Enrichment Problem	Answer
Chase has 25 marbles. He gave Trevor 9 marbles and Kirk 6 marbles. How many marbles does Chase have now? Write 2 number sentences to show how many marbles Chase has now.	10; 25 – 9 = 16 marbles and 16 – 6 = 10 marbles or 9 + 6 = 15 marbles and 25 – 15 = 10 marbles

- **Adding Three Two-Digit Numbers With a Sum Greater Than 100**

English Learners

materials: penny, nickel, and dime

- **compare:** see or show how things are alike or different
- Remind the child that when sorting objects, the first step is to compare the items, to see how they are alike or different.
- Ask the child to compare the coins; ask him or her to tell you two ways the coins are alike and two ways they are different. (possible answers: alike—all are circular, metal, money, have a face on one side; different—penny is not the same color, faces, sizes, reverse sides, values are different)
- Have the child use a complete sentence to explain *compare.*

Spanish Cognate: compare compare

Inclusion Students

materials: index cards (25 money cards) created for the lesson, one 100-flat, 20 10-sticks, and 30 unit cubes

- Ask the child to select three cards; ask him or her to show the values using 10-sticks and unit cubes.
- Write the three numbers vertically on the board as an addition problem.
- Ask the child to use the base ten blocks to add the three numbers, trading unit cubes for a 10 stick, if necessary, and 10-sticks for a 100-flat, if necessary.
- Reflect the steps on the addition problem on the board; compare the results.
- Repeat with three different cards.

At-Risk Students

materials: index cards (25 money cards) created for the lesson

- Ask the child to select three of the cards; ask him or her to write the addition problem vertically on a sheet of paper, being sure to write the tens and ones in straight columns.
- Ask the child to find the sum, assisting him or her as needed.
- Repeat with three different cards.

Advanced Learners

materials: index cards (25 money cards) created for the lesson

- Ask the child to select four of the cards; ask him or her to write the addition problem vertically on a sheet of paper, being sure to write the tens and ones in straight columns.
- Ask the child to find the sum.
- Repeat with four different cards.

- **Subtracting 4 Facts**
- **Drawing a Picture to Solve a Problem**

English Learners

materials: two handfuls of color tiles

- Write *estimate* on the board.
- **estimate:** to make a good guess about value or measurement based on given information
- Explain that the word you have written on the board may be said in two different ways, depending on how the word is used.
- Remind the child that when it is used to show an action, the word is pronounced *est´-e-mate,* meaning to make a reasonable guess.
- Explain that when the word means the guess itself, it is pronounced *est´-e-mit.*
- Ask the child to estimate the number of tiles and then count them to see how close he or she came to the actual number.
- Ask the child to explain the difference between *est´-e-mate* and *est´-e-mit.*

Spanish Cognates: estimate (v) estimar
estimate (n) estimación

Inclusion Students

The child will practice subtracting 4 facts.

materials: Teacher Fact Cards (Set L, Subtracting 4 Facts) and set of Learning Wrap-Ups, #4

- Show the Teacher Fact Cards in random order; ask the child to say the answers.
- Have the child say the subtracting 4 facts aloud as he or she wraps the board.

At-Risk Students

materials: Problem-Solving Worksheet 80A and Performance Task Worksheet 80B

- Review with the child the problem-solving strategies and Problem-Solving Worksheet 80A and Performance Task Worksheet 80B.

Advanced Learners

- Extension and Enrichment problem for Problem-Solving Worksheet 80A (Extension and Enrichment Student Problem [page 18] at www.SaxonMathResources.com)

Extension and Enrichment Problem	Answer
Eight people can ride in each van. Five people can ride in each car. Thirty-six people are going on a trip together. They want to take as many vans as possible, but the vans must be full. The rest of the people will ride in a car. Draw a picture to show how many vans and cars they will need to take. The vans cost $100 to rent, and the cars cost $70 to rent. How much money will they pay altogether to rent the vans and cars?	4 vans with 8 people each and 1 car with 4 people; $470.00

- **Cutting a Geometric Shape Apart and Making a New Shape**

English Learners

materials: several playing cards and a coin

- **flip:** to turn over a figure from front to back or back to front
- Show the child the cards face down and then being turned over; explain that when you turn the card over, you flip it.
- Flip the coin in the air; explain that the coin is turning front to back and back to front many times; say that since the coin is flipping so many times and so quickly, it is necessary to guess whether it will land face up or face down.
- Explain that food may be flipped so that it cooks the same amount on both sides.
- Ask the child what kind of food he or she has seen being flipped. (Possible answers: an egg, a pancake, a hamburger patty, etc.)
- Have the child use a complete sentence to explain *flip*.

Inclusion Students

materials: pattern blocks

- Ask the child to create each of the following shapes:
  1. a triangle with two different-color blocks (trapezoid and triangle)
  2. a trapezoid with two different-color blocks (blue parallelogram and triangle)
  3. a parallelogram with two of the same color blocks (triangles)

At-Risk Students

materials: pattern blocks

- Ask the child to create each of the following shapes:
  1. a triangle with two different-color blocks (trapezoid and triangle)
  2. a trapezoid with two different-color blocks (blue parallelogram and triangle)
  3. a parallelogram with two of the same color blocks (triangles)
  4. a hexagon with three different-color blocks (parallelogram, trapezoid, and triangle)
  5. a square with four of the same color blocks (squares)
  6. a triangle with four of the same color blocks (triangles)

Advanced Learners

materials: 3 sets of tangram pieces (yellow, green, and blue) and yellow, green, and blue crayons

- Ask the child to make as many different rectangles and squares as possible using one yellow, one green, and one blue piece for each shape; ask him or her to use the crayons to draw a picture of each rectangle or square created.

• Using Comparison Symbols (>, <, and =)

English Learners

- On the board, write > with *greater than* below it, < with *less than* below it, and = with *equal to* below it.
- **comparison symbols:** symbols that show whether a number is greater than (>), less than (<), or equal to (=) another number
- **equal to:** having the same value
- Review **greater than:** more than
- **less than:** not as many as
- Review with the child the comparison symbols and the terms.
- Write number pairs on the board leaving a space between the numbers and ask the child to write the correct comparison symbols.
- Erase the words under the three symbols; have the child identify the three comparison symbols and explain to you what each means.

Spanish Cognate: comparison symbols símbolos de comparación

Inclusion Students

materials: 10 index cards: 8 separate cards with the numbers: 18, 5, 25, 12, 7, 11, 23, 31; and 2 cards, one with > and the words *greater than* below it and one with < and the words *less than* below it.

- Have the child select two number cards and place them on the table with a space between the cards.
- Ask the child if the first number is greater than or less than the second number and to place the correct comparison symbol card between them.
- Repeat with other cards.

At-Risk Students

- Math Center Activity 69

Objectives	Materials for Four Children	Teacher Directions
• Using Comparison Symbols (<, >, and =)	• 4 Learning Palettes • 4 Learning Palette Card 32	Ask each child to match the problems and answers on Learning Palette Card 32.

Advanced Learners

- Learning Palette Enrichment Card E12, Using Comparison Symbols (<, >, =)

- **Reading and Drawing a Pictograph With a Scale of 2**

English Learners

- On the board, draw a simple pictograph with the following: 3 children's names followed by circles to indicate the number of cookies eaten; for the 1st child show 1 circle; for the 2nd child show 2 circles; and for the 3rd child show 1½ circles. Under the pictograph, show that a circle = 2 cookies.
- Review that in the pictograph created in the lesson, the scale was one happy face equal to two children; on the graph itself, a half of a happy face would mean one child.
- **scale:** the number shown in the key of a graph or map to show the value of a picture or distance
- Explain to the child that the word *scale* has several different meanings. The child may have heard the word used in other ways before, such as a tool for measuring weight, the scale of a fish, or a musical scale.
- Review that in the lesson, a scale was used to show the value of a picture on a pictograph; generally this is shown in a key at the bottom of the graph.
- Explain that when reading a graph, it is important to look for a key that shows what the scale for that pictograph is, because it may vary from graph to graph; note that if there is no scale shown, each picture equals one.
- Ask the child to point to the key for the pictograph on the board and tell you what it shows.
- Ask him or her to tell you how many cookies each child ate.

Spanish Cognate: scale escala

Inclusion Students

- If the child has trouble understanding the concept of a scale of 2, review the different parts of the pictographs on Lesson Worksheet 82 and what they mean.
- Point out that the Meeting Board thermometer uses a scale of 2; each line indicates 2°F.
- On the board, draw a simple pictograph with the following: 3 children's names followed by circles to indicate the number of cookies eaten; for the 1st child show 1 circle; for the 2nd child show 2 circles; and for the 3rd child show 1½ circles. Under the pictograph, show that a circle = 2 cookies.
- Ask the child to point to the key for the pictograph and tell you what it shows.
- Ask him or her to tell you how many cookies each child ate.

At-Risk Students

- Math Center Activity 70

Objectives	Materials for Four Children	Teacher Directions
• Reading a Pictograph With a Scale of 2 • Reading a Tally Chart	• 4 Learning Palettes • 4 Learning Palette Card 33	Ask each child to match the problems and answers on Learning Palette Card 33.

Advanced Learners

- Learning Palette Enrichment Card E13, Reading Pictographs with Scales of 5 and 10

- **Writing a Fraction to Show a Part of a Set**
- **Picturing a Fractional Part of a Set**

English Learners

materials: Math Folder (Fractions) and color tiles (2 red, 1 blue, and 1 yellow)

- **fractional part of a set:** a number that tells the number of parts of a group
- Remind the child that a set is a group.
- Remind the child that previously he or she has worked with the fractional part of a whole, a whole being one single item such as a circle.
- On the board, draw a circle divided in half; label each half ½; also refer the child to the Math Folder (Fractions).
- Explain that a set is made up of similar objects; a set is like a whole since it is one group.
- Further explain that just as you can have a fractional part of a whole, you can have a fractional part of a set.
- On the board, draw a rectangle with two circles inside and label each circle ½.
- Explain that this is a set of circles, with each circle being one half of the set.
- Place the four color tiles in front of the child, saying that they represent one set of color tiles.
- Tell the child that the two red color tiles are one half of the set.
- Ask the child to say what part of the set the blue tile is and what part the yellow tile is.
- Have the child use a complete sentence to explain *fractional part of a set.*

Inclusion Students

materials: color tiles (2 red, 2 blue, 1 green, and 1 yellow)

- Place 1 yellow and 2 red color tiles in front of the child.
- Ask him or her how many tiles make up this set.
- Ask the child what fraction of tiles are red; then ask what fraction of tiles are yellow.
- Place 1 yellow, 1 blue, and 2 red color tiles in front of the child.
- Ask him or her how many tiles make up this set.
- Ask what fraction of the tiles are red; ... yellow; ... blue; ... not red.
- Repeat using all six color tiles.

At-Risk Students

- Math Center Activity 71

Objectives	Materials for Four Children	Teacher Directions
• Identifying the Fractional Part of a Set	• 4 Learning Palettes • 4 Learning Palette Card 34	Ask each child to match the problems and answers on Learning Palette Card 34.

Advanced Learners

- Learning Palette Enrichment Card E14, Identifying the Fractional Part of a Set

• Writing a Number in Expanded Form

English Learners

- Write the number 648 on the board.
- **expanded form:** a way of showing a number by showing the value of each digit
- Explain to the child that the word expand means to get bigger, and when writing a number in expanded form, it means to write it in a longer way.
- Review that when writing an expanded number, the number is broken down into its parts—hundreds, tens, and ones.
- Beside 648, write = 600 + 40 + 8.
- Write 207 = on the board and ask the child to write the expanded form; explain, if needed, that there are no tens in the number.
- Ask the child to explain to you using a complete sentence what the term is called when numbers are written showing the value of each digit.

Inclusion Students

materials: 15 index cards: on 5 cards write the numbers: 89, 149, 321, 525, 764; on 5 cards write the expanded form of these numbers; and on the remaining 5 cards show the numbers pictorially, using a large square for 100, a tower for 10, and a small square for 1

- Ask the child to match the number to its expanded form and to its picture.

At-Risk Students

- Math Center Activity 72

Objectives	Materials for Four Children	Teacher Directions
• Identifying a Number Written in Expanded Form	• 4 Learning Palettes • 4 Learning Palette Card 35	Ask each child to match the problems and answers on Learning Palette Card 35.

- Math Center Activity 73

Objectives	Materials for Four Children	Teacher Directions
• Identifying a Number Written in Expanded Form Using Words	• 4 Learning Palettes • 4 Learning Palette Card 36	Ask each child to match the problems and answers on Learning Palette Card 36.

Advanced Learners

- Online Activity, Numbers in Expanded Form
  (Saxon Activity Center is available at www.SaxonMathActivities.com.)

- **Subtracting 5 Facts**

English Learners

- Write the following fact family on the board: 11 – 5 = 6, 5 + 6 = 11, 6 + 5 = 11, 11 – 6 = 5
- Review **plus:** a word used between numbers that means to add the numbers together
- Review **minus:** a word used between numbers that means to subtract the second number from the first number
- Review **equal to:** having the same value
- Ask the child to write the symbols for plus, minus, and equal to.
- Review that the equations that are written on the board show a fact family.
- Ask the child to read the four equations.

Spanish Cognate: minus menos

Inclusion Students

materials: Teacher Fact Cards (Set M, Subtracting 5 Facts) and set of Learning Wrap-Ups, #5

- Show the Teacher Fact Cards in random order; ask the child to say the answers.
- Have the child say the subtracting 5 facts aloud as he or she wraps the board.

At-Risk Students

- Math Center Activity 74

Objectives	Materials for Four Children	Teacher Directions
• Subtraction Facts: Subtracting 5	• 4 sets of Wrap-Ups, #5	Allow the children to use Wrap-Up 5 to practice subtracting 5 from a number.

Advanced Learners

Journal Writing (Overview 9)

- Have the child write a paragraph to explain what a Wrap-Up is and how to use it.

- **Covering Designs With Tangram Pieces**

English Learners

materials: tangram set and a copy of one square from Teacher Master 85-2 (back of Lesson 85-2)

- **tangram:** a square that has been divided into seven shapes
- Explain that a tangram is used to practice putting shapes together in different ways.
- Have the child cover the design using the tangram pieces.
- Ask the child what the finished shape is called.
- Have the child use a complete sentence to explain what a *tangram* is.

Inclusion Students

materials: tangram set and a copy of one square from Teacher Master 85-2 (back of Lesson 85-2)

- Have the child cover the design using the tangram pieces.
- Ask the child what the finished shape is called.
- Ask him or her to identify the shape of each tangram piece.

At-Risk Students

- Math Center Activity 75

Objectives	Materials for Four Children	Teacher Directions
• Covering Designs Using Tangrams	• 8 sets of tangrams (2 sets of each color) • Center 75 Masters A and B *Make 4 copies of each.*	Ask each child to use different-color tangram sets to cover each design on Center 75 Masters A and B.

Advanced Learners

materials: tangram set

- Give the child all the pieces (5) except the two large triangles.
- Ask him or her to create a square using all 5 pieces.

- **Writing Money Amounts Using Dollar Signs and Cent Symbols**

English Learners

- Write two money amounts on the board, one with a dollar sign and decimal point, and the other with a cent symbol.
- **cent symbol:** the symbol for the word *cent* or *cents*
- **decimal point:** the symbol that separates the dollars from the cents when writing an amount of money
- **dollar sign:** the symbol for an amount of U.S. money equal to 100 cents
- Point to the money amounts and review with the child cent symbol, decimal point, and the dollar sign.
- Write the following money amounts on the board and ask the child to read them: 61¢, $2.18, 45¢, and $9.55.
- Have the child identify the dollar signs, cent symbols, and the decimal points.

Spanish Cognates:	cent symbol	símbolo de centavos
	decimal point	punto decimal
	dollar sign	símbolo de dólar

Inclusion Students

materials: 12 index cards showing the following amounts: 112¢, $1.12, 68¢, $0.68, 45¢, $0.45, $4.50, 450¢, $6.80, 680¢, $3.94, 394¢

- Ask the child to match the pairs of cards that show the same amount of money.

At-Risk Students

materials: 10 dimes, 10 nickels, and 20 pennies

- Place some coins in front of the child; have him or her count the money and write the amount in two ways.
- Repeat, including some amounts that are more than one dollar.

Advanced Learners

Journal Writing (Overview 9)

- Ask the child to think of something that he or she would like to buy that costs less than $20.00.
- Have the child write a paragraph about what he or she wants to buy, what it costs, and why he or she wants to buy it.

Guide to Differentiated Instruction

- **Subtracting Two-Digit Numbers Using Dimes and Pennies (Part 1)**

English Learners

- **left:** a word used in subtraction problems to show what remains
- Point out to the child the word left has several meanings; review that one meaning he or she has learned is the direction that is the opposite of right.
- Explain that in subtraction, left describes what is remaining after you have used what is needed—it is the extra amount.
- Explain that in the lesson, the remaining amount referred to money, but the word left can refer to other things as well.
- As an example, say: You want to create a design using pattern blocks. There are 15 pattern blocks available to use, but the design only needs 12. How many pattern blocks are left?
- Write the number sentence 15 – 12 = on the board.
- Ask the child to complete the subtraction and tell you what the answer means.
- Have the child use a complete sentence to explain *left.*

Inclusion Students

materials: pennies

- Have the child create and act out a story problem that includes subtracting one-digit numbers, using pennies and the appropriate vocabulary (difference, left, etc.).

At-Risk Students

materials: dimes, pennies, and Workmat 28

- Have the child create and act out a story problem that includes subtracting two-digit numbers, using dimes, pennies, and the appropriate vocabulary (difference, left, etc.).

Advanced Learners

- Extension and Enrichment problem for Guided Class Practice 87, problem 6 (Extension and Enrichment Student Problem [page 20] at www.SaxonMathResources.com)

Extension and Enrichment Problem	Answer
Draw 6 circles and color 1/3 of them. How many of the circles did you color?	6 circles; 2 shaded; 2 circles

• Subtracting Two-Digit Numbers Using Dimes and Pennies (Part 2)

English Learners

materials: the Meeting Board Problem of the Day for Lesson 88 as reference

- Review **dozen:** a set of twelve (Lesson 88 Problem of the Day)
- Review **half dozen:** a set of six (Lesson 88 Problem of the Day)
- Remind the child that the word *dozen* has been used before to describe things that come in a set of twelve, such as eggs, doughnuts, or roses.
- Review the Problem of the Day; ask how many dimes Ryan has and the total amount of money this is; ask how many nickels Anna has and the total amount of money this is.
- Ask the child what a set of 12 is called and what a set of 6 is called.

Spanish Cognate: dozen docena

Inclusion Students

materials: Workmat 28, a cup of 10 dimes, and a cup of 20 pennies

- Write the following subtraction problem vertically on the board: 32¢ – 12¢; write the words *Dimes* and *Pennies* over the appropriate columns.
- Ask the child to tell you how many dimes and pennies are needed to show 32¢ and to show this amount on the workmat; ask him or her to take away 12¢ to solve the problem; show the answer on the board.
- Repeat with 68¢ – 25¢.

At-Risk Students

materials: Workmat 28, a cup of 10 dimes, and a cup of 20 pennies

- Write the following subtraction problem vertically on the board: 32¢ – 18¢; write the words *Dimes* and *Pennies* over the appropriate columns.
- Ask the child to tell you how many dimes and pennies are needed to show 32¢ and to show this amount on the workmat; ask the child what he or she needs to do to take away 18¢ (trade a dime for 10 pennies); ask the child to finish solving the problem; show the answer on the board.
- Repeat with: 72¢ – 21¢, 68¢ – 25¢, 44¢ – 38¢.

Advanced Learners

- Extension and Enrichment problem for Guided Class Practice 88, problem 1 (Extension and Enrichment Student Problem [page 20] at www.SaxonMathResources.com)

Extension and Enrichment Problem	Answer
Use the list to find how many more postcards the child with the most postcards has than the child with the fewest. Write a number sentence to show this. Samira 118 Allison 224 Franklin 96 Morten 191	224 postcards – 96 postcards = 128 postcards

• Subtracting Two-Digit Numbers (Part 1)

English Learners

- **only:** no one or nothing more; nothing else
- On the board write *Sally had only dimes and pennies.*
- Explain that in this sentence *only* means that these were the coins she had, there were no others.
- Have the child use a complete sentence to explain *only.*

Inclusion Students

materials: Center 76 Master, 2 blue index cards with the following money amounts: 72¢, 93¢; and 2 yellow index cards with the following money amounts: 15¢, 37¢; dimes and pennies

- Ask the child to choose a blue and a yellow card and record the amounts on Center 76 Master.
- Ask him or her to find the difference using the coins and the subtraction algorithm.
- Repeat the activity using the other blue and yellow cards.

At-Risk Students

- Math Center Activity 76

Objectives	Materials for Four Children	Teacher Directions
• Subtracting Two-Digit Numbers With Regrouping Using Dimes and Pennies	• 4 blue index cards with one of the following numbers on each card: 80¢, 72¢, 61¢, and 93¢ • 4 yellow index cards with one of the following numbers on each card: 15¢, 26¢, 37¢, and 49¢ • 2 Lesson 28 Work Mats • 2 cups of 10 dimes • 2 cups of 20 pennies • Center 76 Master *Make 1 copy for each child in class.*	Pair the children. Ask one partner to choose a blue card and the other a yellow card. Ask the partners to record the amounts from their cards on Center 76 Master and then find the difference using the subtraction algorithm. Ask the child with the blue card to count out the dimes and pennies to show the amount written on his/her card. Ask the child with the yellow card to take away the amount shown on his/her card, trading a dime for 10 pennies if needed. Ask the children to repeat the activity by choosing another blue card and yellow card. Repeat one more time.

- **Subtracting 6 Facts**
- **Solving a Problem by Guessing and Checking**

English Learners

materials: Math Folder (Months of the Year and hundred number chart)

- Ask the child to look at the *Months of the Year* chart (table) in the Math Folder.
- **chart:** a general term for a way to show information
- Explain that there are different ways to present information and that the best way depends on the type of information to be shown; a chart should be as simple as possible as long as it displays everything that needs to be shown.
- Explain that types of charts include graphs, which use symbols, points, bars and lines; tables, which present values in columns and rows; and diagrams, which have drawings.
- Ask the child to point out charts in the classroom.
- Ask the child what the table on the back of the Math Folder is called. (hundred number chart)

Inclusion Students

The child will practice the subtracting 6 facts.

materials: Teacher Fact Cards (Set N, Subtracting 6 Facts) and set of Learning Wrap-Ups, #6

- Show the Teacher Fact Cards in random order; ask the child to say the answers.
- Have the child say the subtracting 6 facts aloud as he or she wraps the board.

At-Risk Students

materials: Problem-Solving Worksheet 90A and Performance Task Worksheet 90B

- Review with the child the problem-solving strategies and Problem-Solving Worksheet 90A and Performance Task Worksheet 90B.

Advanced Learners

- Extension and Enrichment problem for Problem-Solving Worksheet 90A (Extension and Enrichment Student Problem [page 21] at www.SaxonMathResources.com)

Extension and Enrichment Problem	Answer
The bookstore is having a sale. Hardback books cost 82¢. Paperback books cost 28¢. Magazines cost 14¢. Bookmarks cost 7¢. Sepehr began with $2.00 and selected 2 hardback books. Show the different things he can buy with the remaining money, so that he will receive as little change as possible. How much change will he receive?	1PB, 1BM; 2M, 1BM; 1M, 3BM; or 5BM; 1¢

• Covering the Same Design in Different Ways Using Tangram Pieces

English Learners

materials: tangram set

- Review **tangram:** a square that has been divided into seven shapes
- Ask the child to identify each shape in the tangram set.
- Ask him or her to use a complete sentence to explain what a *tangram* set is.

Inclusion Students

materials: tangram set and Lesson Worksheet 90-2 (from class)

- Have the child identify the different geometric shapes in a tangram set.
- Ask the child to cover each design on the worksheet.

At-Risk Students

materials: tangram set and Lesson Worksheet 90-2 (from class)

- Ask the child to cover each design on the worksheet in two or three different ways.

Advanced Learners

materials: tangram set

- Ask the child to try to assemble a large square using all 7 pieces. (solution on Teacher Master 85-2.)

• Subtracting Two-Digit Numbers (Part 2)

English Learners

materials: Workmat 28, 3 dimes, and 11 pennies

- Write the following subtraction problem vertically on the board: 31¢ – 25¢.
- **step:** one stage of an activity
- Explain that there are several meanings of the word *step;* the child may be familiar with meanings such as the motion used or distance moved in walking; or the meaning stair; but as described in the lesson, the whole subtraction process is made up of individual parts or steps.
- Ask the child to solve the problem shown using dimes and pennies; remind the child of the three steps or parts of the subtraction process discussed in the lesson:
  Step 1. Check to see if there are enough pennies to subtract from;
  Step 2. Trade a dime for ten pennies, if necessary; and
  Step 3. Complete the subtraction of pennies and dimes.
- Have the child use a complete sentence to explain *step.*

Inclusion Students

materials: Workmat 28, 10 dimes, and 20 pennies

- Write the following subtraction problem vertically on the board: 43¢ – 18¢.
- Ask the child to show 43¢ on the workmat using dimes and pennies.
- Review the steps of the subtraction process: check if there are enough pennies to subtract from; trade a dime for ten pennies, if necessary; and complete the subtraction of pennies and dimes.
- Have the child complete the subtraction with coins as you show the steps on the board.
- Write the following subtraction problem vertically on the board: 85¢ – 67¢ and ask the child to tell you the steps needed to solve the problem; complete the subtraction on the board as he or she solves the problem with coins.

At-Risk Students

- Math Center Activity 78

Objectives	Materials for Four Children	Teacher Directions
• Subtracting Two-Digit Numbers	• 4 Learning Palettes • 4 Learning Palette Card 37	Ask each child to match the problems and answers on Learning Palette Card 37.

Advanced Learners

Journal Writing (Overview 10)

- Ask the child to pretend that he or she is going shopping. Have the child write a some, some went away story about how much money he or she had and how much was spent.

- **Writing Number Sentences to Show Equal Groups**
- **Multiplying by 10**

English Learners

materials: 30 blue color tiles, separated into 3 groups of 10

- **equal groups:** sets of objects having the same number or value
- **multiplication:** an operation to combine equal groups
- Explain that there are three groups of ten color tiles; on the board write 3 groups of 10 = 30, 3 × 10 = 30, and as a vertical problem write 10 × 3 = 30; review that the number sentence and problem are the same as writing the equal groups sentence, and each can be read as 3 groups of 10.
- Explain that when the class has been using the counting strips during each Meeting, they have been counting by equal groups; describe multiplication as a quick way to add equal groups.
- Have the child use a complete sentence to explain *equal groups* and *multiplication.*

Spanish Cognates:

equal groups	grupos iguales
multiplication	multiplicación

Inclusion Students

materials: the Meeting Board counting strips and 30 blue color tiles, separated into 3 groups of 10

- Refer to the 10's counting strip as you count by 10's to 30.
- Show that there are 10 tiles in each group; point to them as you again count by 10's to 30.
- Explain that counting by groups is a much faster way to count to 30 than by counting individual pieces.
- Write 3 groups of 10 = 30 on the board; say that since there are three groups of ten color tiles, you can write the sentence shown on the board.
- Write 3 × 10 = 30 on the board below the equal groups sentence; review that the number sentence is the same as writing the equal groups sentence.
- Give the child 20 of the tiles and ask him or her to create 2 groups of 10; ask the child to tell you what the equal groups sentence and number sentence should be.

At-Risk Students

materials: the Meeting Board counting strips and ten 10-sticks

- Refer to the 10's counting strip as you count by 10's to 100.
- Review that each stick represents a group of 10; point to them as you again count by 10's to 100.
- Write 10 groups of 10 = 100 on the board; say that since there are 10 groups of 10, you can write the sentence shown on the board.
- Write 10 × 10 = 100 both horizontally and as a vertical problem on the board below the equal groups sentence; review that the number sentence and problem are the same as writing the equal groups sentence and that both can be read as 10 groups of 10.
- Give the child 4 of the 10-sticks; ask him or her to write the equal groups sentence, number sentence, and number problem the sticks show.
- Repeat with 7 of the 10-sticks and 3 of the 10-sticks.

- **Counting Quarters**

English Learners

materials: 4 quarters

- **quarter:** a U.S. coin worth 25 cents; one fourth
- Draw a rectangle on the board and divide it with a vertical line and horizontal line into four equal parts; write ¼ in each part.
- Explain that the word *quarter* is a general word that means one fourth of something.
- Draw a second rectangle with the same size and divisions as the first; write 25¢ in each part.
- Review that a quarter is the U.S. coin that is equal to one fourth of a dollar; and that a quarter is worth 25 cents.
- Using the quarters, count with the child: 25 cents, 50 cents, 75 cents, one dollar.
- Ask the child to look at the heads side of a state* quarter and read aloud what is written under the head. [*on a quarter minted prior to 1999 this appears under the eagle on the reverse side] (quarter dollar)

Spanish Cognate: quarter cuarto

Inclusion Students

materials: 12 quarters

- Draw a rectangle on the board; divide the rectangle into fourths and write 25¢ in each part.
- Review that a quarter is one fourth of a dollar and that four quarters equal one dollar.
- Ask the child to divide the coins into groups that each equal one dollar; ask him or her how many dollars there are.

At-Risk Students

materials: the Meeting Board counting strips and 12 quarters

- Draw a rectangle on the board; divide the rectangle into fourths and write 25¢ in each part.
- Review that a quarter is one fourth of a dollar and that four quarters equal one dollar.
- Ask the child to divide the coins into groups that each equal a dollar; ask him or her how many dollars there are.
- Remind the child that *one dollar* is another way of saying 100 cents.
- Count the amount of money with the child, saying: 25 cents, 50 cents, 75 cents, 1 dollar, 1 dollar and 25 cents, ... and so on to three dollars.
- Point to the counting by 25's counting strip and count with the child from 0 to 300.

Advanced Learners

- Learning Palette Enrichment Card E15, Identifying Sets of Coins with an Equivalent Value

- **Rounding to the Nearest Ten**

English Learners

materials: classroom number line

- **rounding:** changing a number to show the nearest ten, hundred, and so on
- Remind the child that when rounding a number to the nearest ten, we find the ten that the number is closest to.
- Ask the child which two tens the number 24 is between and which it is closer to; he or she may refer to the number line, if necessary.
- Explain that since 24 is closer to 20, it will be rounded down to 20.
- Repeat with: 41, 18, 57.
- Review with the child that when the number being rounded ends in 5, such as 15, the number is always rounded up to the higher ten.
- Ask the child which two tens the number 15 is between and what number it should be rounded to.
- Repeat with: 35, 85.
- Have the child use a complete sentence to explain the process of *rounding.*

Spanish Cognate: rounding redondeo

Inclusion Students

materials: all pink number cards and 10 randomly selected yellow number cards (from Lesson 94)

- Ask the child to place the pink cards on the floor in order from least to greatest.
- Review that when rounding a number to the nearest ten, we name the ten the number is closest to.
- Ask the child to place each yellow number card below the pink card (multiple of 10) the number is closest to; remind the child to round up if the number on the yellow card ends in 5.
- Continue with additional yellow cards, if desired.

At-Risk Students

- Math Center Activity 80

Objectives	Materials for Four Children	Teacher Directions
• Rounding a Number to the Nearest Ten	• 4 Learning Palettes • 4 Learning Palette Card 38	Ask each child to match the problems and answers on Learning Palette Card 38.

Advanced Learners

Journal Writing (Overview 10)

- Have the child write a paragraph about how he or she would explain to a first-grader how to round numbers to the nearest 10.

- **Subtracting 7 Facts**

English Learners

materials: set of Learning Wrap-Ups and a copy of Lesson Worksheet 95-1A

- **notch:** a small gap; a small cut on an edge or surface
- Point to the indentations along the sides of a plastic Wrap-Up, explaining that these gaps or cuts are called notches.
- Remind the child that when wrapping the string around the board, the string is placed in the notch beside each number
- Remind the child that when completing a Lesson Worksheet with a picture of a Wrap-Up, a straight line should be drawn from the notch at the left of each number to the notch at the right of the answer.
- Ask the child what the gap or cut on the edge of a Wrap-Up is called.

Inclusion Students

materials: Teacher Fact Cards (Set O, Subtracting 7 Facts) and set of Learning Wrap-Ups, #7

- Show the Teacher Fact Cards in random order; ask the child to say the answers.
- Have the child say the subtracting 7 facts aloud as he or she wraps the board.

At-Risk Students

- Math Center Activity 81

Objectives	Materials for Four Children	Teacher Directions
• Subtraction Facts: Subtracting 7	• 4 sets of Wrap-Ups, #7	Allow the children to use Wrap-Up 7 to practice subtracting 7 from a number.

Advanced Learners

materials: two sets of Learning Wrap-Ups, #7

This activity is for two children.

- Have Wrap-Up Races: ask the children to prepare to wrap the #7 board; when they are ready, start the race.
- The winner of each race is the child whose board is correctly wrapped first; the overall winner wins two out of three.

- **Estimating and Counting Large Collections**

English Learners

- **collection:** a group of like or similar things
- Remind the child that during the lesson, the class used bags of items, and the objects in each bag (such as beans, paper clips, and so on) were alike; these were collections.
- Explain that the word *collection* can also be used to describe groups of things that are similar, like the paintings in a museum or the songs on a CD.
- Explain that many people like to have their own collection of things they especially like, such as stuffed animals, game cards, dolls, books, butterflies, stamps, and so on.
- Ask the child if he or she has ever started a collection, and if so, of what.
- Have the child use a complete sentence to explain *collection.*

Spanish Cognate: collection colección

Inclusion Students

materials: bag of 65 small objects (modified from Lesson 95-2) and 6 small paper cups

- Ask the child to estimate the number of objects in the bag and to write the estimate on a piece of paper.
- Have the child count the objects in the bag, using the paper cups to put them in groups of 10.
- Ask him or her how many objects were left that could not be put into a group of 10.
- Ask the child what the actual number of objects in the bag was and how close his or her estimate was.

At-Risk Students

- Math Center Activity 82

Objectives	Materials for Four Children	Teacher Directions
• Estimating and Counting Large Collections	• bags of 200–400 objects (from Lesson 95-2) • 40 small paper cups • children's Lesson Worksheet 95-2 (from Lesson 95-2)	Pair the children. Ask the partners to work together to estimate and count the objects from a collection that they did not estimate and count during Lesson 95-2. The children can use the paper cups to put the objects in groups of 10.

Advanced Learners

Journal Writing (Overview 10)

- Have the child write a paragraph about what he or she would use to fill the estimation jar and why he or she chose that object.

Finding One Half of a Set With an Even Number of Objects

English Learners

- On the board draw two squares, the first evenly divided into fourths and the second unevenly divided into four parts.
- **equal parts:** sections of a whole that are exactly the same size
- Remind the child that when using a fraction to describe part of a whole or part of a set, that the whole or set must be divided into parts that are the same size.
- Fold a sheet of paper in half and then in half again; show the child that the fold lines have divided the paper into four equal parts.
- Have the child use a complete sentence to explain *equal parts.*

Spanish Cognate: equal parts partes iguales

Inclusion Students

materials: 12 blue color tiles

- Fold a sheet of paper in half; ask the child how many equal sections there are and what each section is called.
- Give the child 10 of the color tiles and ask him or her to count them.
- Ask him or her to separate the tiles into two equal groups on the paper.
- Explain that each group is half of the tiles; ask the child how many tiles are in each group.
- Ask the child to recall the doubles fact 5 + 5 = 10 as you write the number sentence on the board; point out that half of the set of 10 is 5 and that there are two sets of 5 in 10.
- Repeat with: 12 color tiles, 6 color tiles.

At-Risk Students

- Math Center Activity 83

Objectives	Materials for Four Children	Teacher Directions
• Writing a Division Story Using Even Numbers	• story paper—1 piece per child • stickers or stamp and stamp pad	Ask each child to write a story about dividing an even number of (objects) into two groups. The child can use stickers, a stamp and stamp pad, or can draw a picture to illustrate his/her story. Ask each child to describe what happened in his/her story.

Advanced Learners

- Ask the child to write a story about dividing an even number of (objects) into two groups.
- Ask him or her to draw a picture to illustrate what took place in the story.

- **Finding One Half of a Set With an Odd Number of Objects**

English Learners

materials: 7 blue color tiles

- **remainder:** the part left after dividing a group of items into equal groups
- Review that the remainder is the piece that is extra when separating an odd number of objects into two equal groups; this is the piece that *remains* apart.
- With the help of the child, divide the tiles into two equal groups; show that the one extra piece cannot be placed in either group because the groups would be unequal.
- Ask the child what this extra piece is called.

Inclusion Students

materials: 11 blue color tiles

- Fold a sheet of paper in half; give the child 11 color tiles and have him or her count them.
- Ask the child to try to separate the tiles into two equal groups on the paper.
- Ask the child what one half of 11 is. (5 R1)
- Review that the extra tile that cannot be placed in a group is called the remainder.
- Ask the child to recall the doubles plus 1 fact 5 + 6 = 11 as you write the number sentence on the board; show that this could also be written 5 + 5 + 1 = 11, showing two groups of 5 and 1 extra.

At-Risk Students

- Math Center Activity 84

Objectives	Materials for Four Children	Teacher Directions
• Writing a Division Story Using Odd Numbers	• story paper—1 piece per child • stickers or stamp and stamp pad	Ask each child to write a story about dividing an odd number of (objects) into two groups. The child can use stickers, a stamp and stamp pad, or can draw a picture to illustrate his/her story. Ask each child to describe what happened in his/her story.

Advanced Learners

- Extension and Enrichment problem for Guided Class Practice 97, problem 1 (Extension and Enrichment Student Problem [page 22] at www.SaxonMathResources.com)

Extension and Enrichment Problem	Answer
Maria brought 2 dozen cookies for her pom class. A half dozen cookies were eaten. How many cookies were left? If there are 9 girls in the class and they divided the cookies that were left to take home, how many cookies did each girl take?	18 cookies; 2 cookies

• Estimating a Sum

English Learners

materials: classroom number line for reference

- **round:** to change a number to show the nearest ten, hundred, and so on
- Explain to the child that the word *round* has several meanings in English; one meaning the child may be familiar with is to describe the curved shape of an object (like a ball) that is smooth and without corners.
- Explain that in mathematics, the term is used like *rounding,* to mean the action of changing a number to the nearest ten; remind the child that numbers ending in 5 are rounded up.
- Using the classroom number line, if necessary, ask the child to round 18, 66, 23, 31, and 45.
- Have the child use a complete sentence to explain how to *round* a number.

Spanish Cognate: round redondear

Inclusion Students

- Write the number sentence 61¢ + 18¢ = on the board.
- Explain that rounding numbers makes it much easier to estimate sums, especially when doing addition in your head.
- Review the process of rounding, if necessary, and ask the child to round each of the numbers in the number sentence.
- Below the number sentence write 60¢ + 20¢ = and ask the child to solve the addition.
- Repeat with: 22¢ + 47¢ = , 18¢ + 21¢ + 32¢ = .

At-Risk Students

- Math Center Activity 85

Objectives	Materials for Four Children	Teacher Directions
• Estimating a Sum	• 4 Learning Palettes • 4 Learning Palette Card 39	Ask each child to match the problems and answers on Learning Palette Card 39.

Advanced Learners

- Extension and Enrichment problem for Guided Class Practice 98, problem 3 (Extension and Enrichment Student Problem [page 23] at www.SaxonMathResources.com)

Extension and Enrichment Problem	Answer
Draw 8 circles and color 6 of them. Is the fraction 6/8 closer to 0 or to 1? Color the rest of the circles. Write the fraction of the colored circles now. What does this fraction equal?	8 circles, 6 colored; closer to 1; 8/8; 1

• Measuring Using Feet and Inches

English Learners

materials: ruler

- **height:** the distance from top to bottom or bottom to top
- Explain that the word height is used to show the measure of how tall something is, like a building, a mountain, or a person; point out that *height* sounds like *high,* which can be used to describe something tall.
- Review that height, like the measure length, is usually shown in feet and inches.
- Measure how tall the child is by having him or her stand against the wall and putting a small mark at the top of his or her head; with the child, measure the distance from the floor to that mark; have the child describe his or her height in feet and inches.
- Have the child use a complete sentence to explain *height.*

Inclusion Students

materials: ruler

- Ask the child to tell you how he or she would measure something longer than 12 inches with a ruler; review that it is necessary to measure in a straight line, to mark any partial measurements before moving the ruler, and to keep track of how many feet have been measured.
- Ask the child to measure the length of a table, desk, or bookshelf, assisting as needed.
- Ask the child to record the measurement in feet and inches on the board.

At-Risk Students

materials: ruler

- Ask the child to tell you how he or she would measure something longer than 12 inches with a ruler; review that it is necessary to measure in a straight line, to mark any partial measurements before moving the ruler, and to keep track of how many feet have been measured.
- Ask the child to measure the length of a table, desk, or bookshelf, assisting as needed.
- Ask the child to record the measurement in feet and inches on the board.
- Ask him or her to measure a vertical length, such as the side of a bookshelf or the side of a solid desk, assisting as needed.
- Ask the child to record the measurement in feet and inches on the board.

Advanced Learners

materials: a ruler (if needed for reference)

- Write the following problem on the board: There are three markers placed end to end in a row. Each marker is 6 inches long. What is the length of the row in inches? What is the length of the row in feet and inches? (Suggest that the child start by drawing a picture of the problem on the board, if necessary.)
- Write the following problem on the board: The side of a square is 10 inches long. What is the total length of the sides of the square in inches. What is the total length of the sides of the square in feet and inches? (Suggest that the child start by drawing a picture of the problem on the board, if necessary.)

- **Subtracting 8 Facts**
- **Making a Table to Solve a Problem**
- **Looking for a Pattern to Solve a Problem**

English Learners

materials: Math Folder

- **table:** a kind of chart; a way to show information in columns and rows
- Explain that *table* has several meanings in English; the child is probably already familiar with its meaning as a piece of furniture.
- Review with the child that a table is also a way to show information; it is a type of chart made up of columns and rows.
- Refer to the Math Folder and point out the tables: Months of the Year, Geometric Solids, and the hundred number chart; point out the columns and rows in each.
- Explain that a table, like a graph, is a way of organizing and showing information so that it is easier to understand and use.
- Have the child use a complete sentence to explain *table*.

Inclusion Students

The child will practice the subtracting 8 facts.

materials: Teacher Fact Cards (Set P, Subtracting 8 Facts) and set of Learning Wrap-Ups, #8

- Show the Teacher Fact Cards in random order; ask the child to say the answers.
- Have the child say the subtracting 8 facts aloud as he or she wraps the board.

At-Risk Students

materials: Problem-Solving Worksheet 100A and Performance Task Worksheet 100B

- Review with the child the problem-solving strategies and Problem-Solving Worksheet 100A and Performance Task Worksheet 100B.

Advanced Learners

- Extension and Enrichment problem for Problem-Solving Worksheet 100A (Extension and Enrichment Student Problem [page 23] at www.SaxonMathResources.com)

Extension and Enrichment Problem	Answer
The children in Ms. Tran's class are selling t-shirts. Each t-shirt costs \$8.00. How much money will the class make if they sell 12 t-shirts? If they lower the price to \$7.00, how much money will they make if they sell 12 t-shirts? Do you think there will be more demand if they sell the t-shirts for \$7.00? How many \$7.00 shirts do they need to sell to make more than they would make selling 12 shirts at \$8.00?	\$96.00; \$84.00; yes; 14 t-shirts

- **Finding the Area of Shapes Using Pattern Blocks**

English Learners

materials: paper shapes (4" × 4" square, 2" × 2" square, 2" × 3" rectangle) and 16 color tiles

- **area:** the number of square units needed to cover a surface
- Review that a line is a single, straight path that has only length, while area is the measurement of a flat surface, such as the inside of a square.
- Ask the child to cover the large square with color tiles; ask how many it took.
- Repeat with the small square and the rectangle.
- Have the child use complete sentences to explain the *area* of the three shapes.

Spanish Cognate: area área

Inclusion Students

materials: paper shapes (4" × 4" square, 2" × 2" square, 2" × 3" rectangle) and 16 color tiles

- Review that a line is a single, straight path that has only length, while area is the measurement of a flat surface, such as the inside of a square.
- Ask the child to cover the large square with color tiles; ask how many it took.
- Repeat with the small square and the rectangle.
- Place the small paper square in one corner of the large paper square; ask the child how many small paper squares would be needed to cover the large square.

At-Risk Students

- Math Center Activity 87

Objectives	Materials for Four Children	Teacher Directions
• Finding the Area of a Shape Using a Nonstandard Unit	• basket of pattern blocks • Center 87 Master *Make 1 copy for each child in class.*	Ask each child to use the pattern blocks to find the area of the shapes on Center 87 Master.

Advanced Learners

materials: pattern blocks (1 hexagon, 2 trapezoids, 6 triangles)

- Have the child write a paragraph explaining why it takes more triangles than trapezoids to cover a hexagon. (Each triangle covers less area, so that it takes more of them to cover the same area.)
- Have the child write a paragraph explaining how area is different from length. (If the child has difficulty, ask him or her to start by explaining the difference between a line and a square.)

Guide to Differentiated Instruction Lesson 101

- **Identifying, Describing, and Comparing Geometric Solids (Cone, Cube, Sphere, Cylinder, Rectangular Prism, and Pyramid)**

English Learners

materials: a variety of cylinders (cans, tubes, etc.) and spheres (baseball, golf ball, basketball, etc.), including the cylinder and sphere from the set of geometric solids

- Show the child a cylinder and a sphere.
- **cylinder:** a geometric solid with congruent circles on each end
- **sphere:** a geometric solid shaped like a globe or ball
- Point to each object as you identify it as a *cylinder* or *sphere.*
- Ask the child to sort the items one-by-one by shape and say either: This is a sphere. or This is a cylinder.

Spanish Cognates: cylinder cilindro
sphere esfera

Inclusion Students

- Review with the child what a sphere is and what a cube is.
- Ask the child to identify items in the classroom that are spheres.
- Ask him or her to identify items in the classroom that are cubes.

At-Risk Students

- Math Center Activity 88

Objectives	Materials for Four Children	Teacher Directions
• Identifying Polygons and Geometric Solids	• 4 Learning Palettes • 4 Learning Palette Card 40	Ask each child to match the problems and answers on Learning Palette Card 40.

Advanced Learners

Journal Writing (Overview 11)

materials: the set of geometric solids

- Have the child choose two geometric solids and write a paragraph explaining two ways they are alike and/or two ways they are different.

Guide to Differentiated Instruction

- **Selecting an Appropriate Tool for Measuring Length**
- **Identifying Metric Units of Length**
- **Measuring and Drawing Line Segments Using Centimeters**

English Learners

materials: ruler

- **centimeter:** a small metric unit used to measure length; cm
- Remind the child that the ruler has two different scales for measuring length.
- On a piece of paper, draw 6 cm and 17 cm line segments.
- Ask the child to measure the lines in centimeters and write the lengths below them.
- Ask the child to draw 8 cm and 12 cm line segments and write the lengths below them.
- Ask the child what three units can be measured with a ruler. (foot, inches, centimeters)

Spanish Cognate: centimeter centímetro

Inclusion Students

The child will select the appropriate tool for measurement.

materials: measuring tools (balance, ruler, meterstick/yardstick, thermometer, demonstration clock, and a measuring cup)

- Remind the child of the advantages of using a ruler for measuring shorter lengths and a meterstick/yardstick for measuring longer lengths.
- Ask the child to select which tool would be used to measure: temperature, the length of a pencil, an hour, the height of a doorway, a cup of milk, time, the weight of a handful of color tiles, 38°F, a minute, the length of a crayon, the weight of three crayons, a cup of water, and the height of a person.

At-Risk Students

The child will measure and draw line segments using centimeters.

materials: ruler

- On a piece of paper, draw 10 cm and 19 cm line segments.
- Ask the child to measure the lines and write the lengths below them in centimeters; ask him or her to measure the segments again to the nearest half inch. (4", 7½")
- Ask the child to draw 5 cm and 14 cm line segments and write the lengths below them using centimeters; ask him or her to measure the segments to the nearest half inch. (2", 5½")

Advanced Learners

- Extension and Enrichment problem for Guided Class Practice 102, problem 1 (Extension and Enrichment Student Problem [page 24] at www.SaxonMathResources.com)

Extension and Enrichment Problem	Answer
A clown at the circus had 30 balloons on Saturday. He gave away 1 dozen balloons. On Sunday he gave away 10 more. How many balloons did he give away altogether? How many balloons did he have left?	22 balloons; 8 balloons

- **Multiplying by 1**
- **Multiplying by 100**

English Learners

- On the board write 10 + 10 + 10 + 10 = 40, 4 groups of 10 = 40, 4 × 10 = 40 (horizontally) and 10 × 4 = 40 (vertically).
- Review **multiplication:** an operation to combine equal groups
- Remind the child that multiplication is a faster way to add; it is similar to using the Meeting Board counting strips.
- Write several of the multiplication by 10 facts (Lesson 92) on the board and ask the child to solve them.
- Have the child use a complete sentence to explain *multiplication.*

Spanish Cognate: multiplication multiplicación

Inclusion Students

The child will practice multiplying by 100.

materials: 10 index cards with the multiplication by 100 facts written on them (1 × 100 = ____, 2 × 100 = ____, 3 × 100 = ____, etc.), 10 index cards with (1 group of 100 = ____, 2 groups of 100 = ____, 3 groups of 100 = ____, etc.) written on them, and 100-flats

- Have the child match the cards and solve the problem using the 100-flats.

At-Risk Students

materials: 12 index cards: nine index cards with a single digit on each (1–9), three index cards with the numbers 1, 10, or 100, and the base ten blocks (unit cubes, 10-sticks, and 100-flats)

- Review with the child that each unit cube is equal to one, each 10-stick is equal to 10, and each 100-flat is equal to 100.
- Ask the child to select a card with a single digit and one of the three index cards with 1, 10, or 100.
- Ask the child to determine which base ten blocks are needed to show the answer if the two numbers are multiplied. (Example: If the child draws a 6 card and a 100 card, he or she will need 6 100-flats because 6 × 100 = 600.)
- Return the cards to the appropriate group and have the child repeat with 2 new cards.
- Repeat the activity as time permits.

Advanced Learners

- Extension and Enrichment problem for Guided Class Practice 103, problem 5 (Extension and Enrichment Student Problem [page 24] at www.SaxonMathResources.com)

Extension and Enrichment Problem	Answer
A new pencil is 7 inches long. How long would 3 new pencils be placed end to end? Give the answer in inches only and then in feet and inches.	21 inches; 1 foot, 9 inches

- **Finding Perimeter**

English Learners

materials: ruler

- **perimeter:** the distance around a shape
- On a piece of paper draw a triangle on one side and a rectangle on the other side; all line segments should be drawn to a whole centimeter.
- Remind the child that finding the perimeter is like a some, some more problem because to find the perimeter of a shape, the lengths of all of its sides must be added together.
- Have the child find the perimeter of the triangle by measuring and recording each side using centimeters and then adding the measurements; remind him or her to show the units.
- Repeat with the rectangle.
- Have the child use a complete sentence to explain *perimeter.*

Spanish Cognate: perimeter perímetro

Inclusion Students

materials: ruler

- On a piece of paper draw a triangle on one side and a rectangle on the other side; all line segments should be drawn to a whole centimeter.
- Remind the child that finding the perimeter is like a some, some more problem because to find the perimeter of a shape, the lengths of all of its sides must be added together.
- Work with the child to find the perimeter of the triangle: measure each side using centimeters, write each measurement beside the line segment, write a vertical addition problem, find the total, and show the units.
- Have the child find the perimeter of the rectangle by measuring all the sides using centimeters; assist as needed.

At-Risk Students

- Math Center Activity 89

Objectives	Materials for Four Children	Teacher Directions
• Finding Perimeter	• 4 Learning Palettes • 4 Learning Palette Card 41	Ask each child to match the problems and answers on Learning Palette Card 41.

Advanced Learners

Journal Writing (Overview 10)

- Have the child write a paragraph to describe how he or she would measure the perimeter of the classroom.
- If the room is a rectangle, ask the child to include in the description why it would not be necessary to measure all four walls to find the perimeter.

- **Subtracting 9 Facts**

English Learners

- Write several subtraction problems on the board.
- Review **difference:** the answer to a subtraction problem
- Ask the child to solve the subtraction problems on the board and to say the difference.
- Ask the child what the answer to a subtraction problem is called.

Spanish Cognate: difference diferencia

Inclusion Students

materials: Teacher Fact Cards (Set Q, Subtracting 9 Facts) and set of Learning Wrap-Ups, #9

- Show the Teacher Fact Cards in random order; ask the child to say the answers.
- Have the child say the subtracting 9 facts aloud as he or she wraps the board.

At-Risk Students

- Math Center Activity 90

Objectives	Materials for Four Children	Teacher Directions
• Subtraction Facts: Subtracting 9	• 4 sets of Wrap-Ups, #9	Allow the children to use Wrap-Up 9 to practice subtracting 9 from a number.

- **Writing Observations From a Graph**

English Learners

materials: the Meeting Board Daily Temperature graph and Our Favorite Ice Cream Flavors graph for reference

- **observation:** a fact based on what is noticed; information gathered by the act of looking carefully
- Remind the child that during The Meeting, the Student of the Day finds and graphs the temperature.
- Remind the child that during the lesson, the class created a graph to determine the class's favorite ice cream flavors.
- Review that both graphs are a collection of information or data.
- Ask the child to make an observation from either graph; if he or she has difficulty, suggest examples like: We have had _____ days over 40°. There are _____ more children that like (chocolate) than (strawberry).
- Have the child use a complete sentence to explain what an *observation* is.

Spanish Cognate: observation observación

Inclusion Students

materials: Our Favorite Ice Cream Flavors graph for reference

- Ask the child to explain the word *observation.*
- Ask him or her to state at least three observations about the graph.
- Encourage the child to state different types of observations, such as: _____ children like (vanilla); _____ more children like (chocolate) than (strawberry); or The least number of children chose _____.

At-Risk Students

materials: Our Favorite Ice Cream Flavor graph

- Ask the child to write the word *Observations* at the top of a sheet of paper.
- Ask him or her to write at least three observations about the graph.
- Encourage the child to write different types of observations, such as: _____ children like (vanilla); _____ more children like (chocolate) than (strawberry); or The least number of children chose _____.

Advanced Learners

- Learning Palette Enrichment Card E16, Comparing Data on a Chart

- **Identifying Activities That Take One Hour, One Minute, and One Second**
- **Telling and Showing Time to the Minute**

English Learners

material: classroom clock with a second hand

- **second:** a very short amount of time; 60 seconds = 1 minute
- Remind the child that second also means the number two position. (Example: first, second, third, etc.)
- Point to the tick marks around the face of the clock and note that each of those small marks indicates one second when the second hand moves from one to the next.
- Show how quickly a second passes by counting the ticks (up to 10) as the second hand passes them.
- Explain that when the second hand has moved once around the clockface, a minute has passed.
- Ask the child to name some things that can be done in a second. (snap fingers, say a number)
- Have the child use a complete sentence to explain *second* as a unit of time.

Spanish Cognate: second segundo

Inclusion Students

The child will tell time to the minute.

materials: student clock

- Write 5:16 on the board; ask the child to say the time and to show the time on the student clock.
- Repeat with: 2:42, 3:15, 9:26.
- On the student clock show 10:27 and ask the child to say the time.
- Repeat with: 4:04, 6:21, 8:53.

At-Risk Students

- Math Center Activity 91

Objectives	Materials for Four Children	Teacher Directions
• Telling Time to the Minute	• 4 Learning Palettes • 4 Learning Palette Card 42	Ask each child to match the problems and answers on Learning Palette Card 42.

Advanced Learners

- Learning Palette Enrichment Card E17, Identifying Times Written Using Words

Guide to Differentiated Instruction

- **Counting Quarters, Dimes, Nickels, and Pennies**
- **Showing Money Amounts Using Quarters, Dimes, Nickels, and Pennies**

English Learners

materials: tangram set and 10 color tiles (use all 4 colors)

- **combination:** a mixture or grouping of items
- Explain that the tangram set is a combination of 7 pieces of various sizes and shapes.
- Explain that the set of color tiles is a combination of different colors.
- Make two groups of 5 color tiles, with each group having several different colors; explain that the two groups show different combinations of colors.
- Have the child use a complete sentence to explain *combination.*

Spanish Cognate: combination combinación

Inclusion Students

The child will count and show money amounts using quarters, dimes, nickels, and pennies.

materials: 3 cups with a varied collection of quarters, dimes, nickels and pennies in each; use fewer than 4 quarters, 5 dimes, 5 nickels, and 5 pennies per cup

- Ask the child to sort the coins from one of the cups and to count the money starting with the highest value coins first; ask him or her to say the total amount.
- Repeat with the other cups.
- On the board write 97¢ and ask the child to show this amount with the fewest coins possible.
- Repeat with $1.32.

At-Risk Students

- Math Center Activity 93

Objectives	Materials for Four Children	Teacher Directions
• Finding the Value of a Set of Coins	• 4 Learning Palettes • 4 Learning Palette Card 43	Ask each child to match the problems and answers on Learning Palette Card 43.

Advanced Learners

- Extend and Challenge CD, Activity 8, Let's Go Shopping
- After the child finishes the activity, use the question(s) on the recording form to evaluate his or her understanding. (Monitoring Student Progress binder, Extend and Challenge CD booklet)

• Identifying Parallel Lines and Line Segments

English Learners

materials: Math Folder (Line Segments) and blue pattern block

- Draw pairs of parallel lines on the board, going horizontally, vertically, and diagonally.
- **parallel:** lines that do not cross and are always the same distance apart
- Describe parallel lines or line segments as being similar to train tracks.
- Explain that parallel lines or line segments may extend in any direction— horizontally, vertically, or obliquely; ask the child to find the parallel line segments in the Math Folder.
- On the blue pattern block, called a parallelogram, show that its opposites sides are both parallel.
- Ask the child to use a complete sentence to explain *parallel.*

Spanish Cognate: parallel paralelas

Inclusion Students

materials: Math Folder (Line Segments), geoboard, and geobands

- On the board, write the word *parallel.*
- Draw pairs of parallel lines on the board going horizontally, vertically, and diagonally.
- Review with the child that parallel lines will always stay the same distance apart and that they will never meet; remind the child that the meaning of *parallel* can be remembered by the two l's, which are parallel line segments, in the middle of the word.
- Show the child the parallel line segments in the Math Folder.
- On the geoboard, show various pairs of line segments, some parallel and some not, and ask the child to say whether the line segments are parallel or not.

At-Risk Students

materials: set of pattern blocks

- On the board, write the word *parallel.*
- Remind the child that the meaning of *parallel* can be remembered by the two l's, which are parallel line segments, in the middle of the word.
- Ask the child to take each of the pattern blocks and point out pairs of parallel lines, if any.
- Remind the child that a parallelogram has two pairs of parallel lines.

Advanced Learners

- Extension and Enrichment problem for Guided Class Practice 108, problem 1 (Extension and Enrichment Student Problem [page 25] at www.SaxonMathResources.com)

Extension and Enrichment Problem	Answer
Miguel has 12 books. He gave 2 books to his sister. He shared the rest equally with his brother. How many books each do Miguel and his brother have?	5 books each

- **Adding Three-Digit Numbers and Money Amounts**
- **Estimating Sums**

English Learners

- On the board write \$1.16 + 1.22 as a vertical addition problem; do not write \$ with 1.22.
- Review **decimal point:** the symbol that separates the dollars from the cents when writing an amount of money
- Review **dollar sign:** the symbol for an amount of U.S. money equal to 100 cents
- Point to the dollar sign in the upper number and the decimal points in both numbers; note that when adding columns of money amounts, it is not necessary to show the dollar sign on more than the top number and the answer.
- Explain that the cents, dimes, decimal points, and dollars should line up vertically in straight columns to make addition easier.
- Ask the child to solve the addition; make sure he or she shows a dollar sign in the answer.
- Have the child use complete sentences to explain both *decimal point* and *dollar sign.*

Spanish Cognates: decimal point — punto de decimal
dollar sign — símbolo de dólar

Inclusion Students

- Write the following addition problem vertically on the board: 116 pennies + 212 pennies.
- Ask the child to solve the problem, beginning with the ones column.
- Then write the numbers vertically as money amounts (\$1.16 + 2.12), review that the \$ is not needed on the lower number; emphasize that the number columns and decimal points must line up.
- Ask the child to re-solve the problem and show the sum as a money amount using the dollar sign and decimal point; explain that both answers show the same amount of money.
- Repeat with 486 pennies + 327 pennies; note that this problem requires carrying ones and tens.

At-Risk Students

- Write the following addition problem vertically on the board: 116 pennies + 212 pennies.
- Ask the child to solve the problem, beginning with the ones column.
- Then write the numbers vertically as money amounts (\$1.16 + 2.12), review that the \$ is not needed on the lower number; emphasize that the number columns and decimal points must line up.
- Ask the child to re-solve the problem and show the sum as a money amount using the dollar sign and decimal point; explain that both answers shows the same amount of money.
- Repeat with 147 pennies + 511 pennies; 486 pennies + 327 pennies; and 446 pennies + 85 pennies; note that the last two problems require carrying ones and tens.

Advanced Learners

- Learning Palette Enrichment Card E18, Adding Three-Digit Numbers

Guide to Differentiated Instruction

- **Multiplying by 5 Facts**
- **Drawing Pictures and Writing Multiplication Number Sentences to Show Equal Groups**
- **Making an Organized List to Solve a Problem**

English Learners

materials: 4 color tiles (1 of each color)

- Review **organize:** arrange in an orderly way
- Explain that the child will create an organized list to show all the possible color combinations of two color tiles.

  Note: A different combination is *not* created when items are placed in a different order; permutations, not combinations, specify order.
- Ask the child to start with the red tile, and write the three combinations that are possible; initials may be used. (RG, RB, RY)
- Ask him or her to continue with the green tile; if the child writes GR, remind him or her that the RG combination has already been used.
- Continue with the remaining tiles. (RG, RB, RY, GB, GY, BY)
- Ask the child to use a complete sentence to explain the meaning of *organize*.

Spanish Cognate: organize organizar

Inclusion Students

materials: 25 color tiles, Teacher Fact Cards (Set R, Multiplying by 5 Facts; plus 2 × 5, 3 × 5, 4 × 5)

- Show the child the cards in random order and ask him or her to say the answers.
- If the child is having difficulty, use only the cards up to 5 × 5 and use the color tiles to explain multiplication by 5's.

At-Risk Students

materials: Problem-Solving Worksheet 110A and Performance Task Worksheet 110B

- Review with the child the problem-solving strategies and Problem-Solving Worksheet 110A and Performance Task Worksheet 110B.

Advanced Learners

- Extension and Enrichment problem for Problem-Solving Worksheet 110A (Extension and Enrichment Student Problem [page 26] at www.SaxonMathResources.com)

Extension and Enrichment Problem	Answer
The sandwich shop offers different types of meats, bread, and cheese. They sell ham, turkey, and roast beef. They have white and rye bread. They have American and Swiss cheese. Show all the different types of sandwiches they can make using 1 meat, 1 bread, and 1 cheese.	WHA, WHS, WTA, WTS, WBA, WBS, RHA, RHS, RTA, RTS, RBA, and RBS

- **Measuring Weight Using Customary Units**

English Learners

materials: bathroom scale and two items to weigh that were used during the lesson

- Show the child the bathroom scale, including the dial or display where weight is read.
- **weight:** the measure of how heavy something is
- **scale:** an instrument used to measure how much something weighs
- Weigh one of the items and remind the child that the scale shows weight in pounds.
- Have the child weigh the second item and tell you its weight; be sure he or she includes the unit pounds; ask if the second item weighs more or less than the first item.
- Ask the child to say who would weigh more, himself or herself or a new baby.
- Ask the child to explain how he or she knows and to use the words weight and scale in the explanation.

Spanish Cognate: scale escala

Inclusion Students

materials: bathroom scale and items to weigh that were used during the lesson

- Have the child weigh the items used in the lesson again, reading the scale, and telling you the weight of each item in pounds.
- Ask the child which item weighs the most and which weighs the least.

At-Risk Students

- Math Center Activity 94

Objectives	Materials for Four Children	Teacher Directions
• Weighing Objects Using Pounds • Ordering Objects from Lightest to Heaviest	• 5 classroom objects between 2 pounds and 15 pounds that will fit easily on the bathroom scale. • bathroom scale • Center 94 Master *Make 1 copy for each child in class.*	Pair the children. Ask each pair to predict the order of the objects by weight from lightest to heaviest. Ask each pair to weigh the objects on the scale and write the names of the objects and their weights on Center 94 Master. Ask each pair to put the objects in order from lightest to heaviest.

Advanced Learners

- Learning Palette Enrichment Card E19, Identifying Equivalents

• Representing and Writing Mixed Numbers (Part 1)

English Learners

- On the board write $3\frac{1}{4}$ and $2\frac{1}{2}$; draw and shade $3\frac{1}{4}$ circles; draw and shade $2\frac{1}{2}$ squares.
- **mixed number:** a number that contains both a whole number and a fraction
- Remind the child that the word *mix* means to combine things or put things together to make something else.
- Explain the mixed numbers on the board, which show a whole number put together with a fraction; show that the shaded drawings represent a mixture of whole shapes plus an additional part, or fraction, of the shape.
- On the board draw and shade $2\frac{2}{3}$ rectangles; ask the child to tell you the mixed number.
- Ask the child to use a complete sentence to explain *mixed number.*

Spanish Cognate: mixed number número mixto

Inclusion Students

materials: pattern blocks

- On the board write: $1\frac{1}{3}$, $4\frac{1}{2}$, $2\frac{1}{2}$, $1\frac{2}{6}$.
- Demonstrate making $1\frac{1}{3}$ using yellow and blue pattern blocks: first show 2 yellow pattern blocks with 3 blue blocks covering the second; review that since it takes three blue pattern blocks to cover one yellow pattern block, each blue pattern block is $\frac{1}{3}$; take away two of the blue pattern blocks and the yellow pattern block on which they sat; explain that the blocks that remain equal $1\frac{1}{3}$.
- Assist the child, as needed, with the following activity; if he or she is having problems visualizing the fractional part, show a whole of the same shape made from the fraction pieces.
- Ask the child to show $4\frac{1}{2}$ using yellow and red pattern blocks. (4 yellow and 1 red)
- Ask the child to show $2\frac{1}{2}$ using blue and green pattern blocks. (2 blue and 1 green)
- Ask the child to show $1\frac{2}{6}$ using yellow and green pattern blocks. (1 yellow and 2 green)

At-Risk Students

- Math Center Activity 95

Objectives	Materials for Four Children	Teacher Directions
• Identifying the Mixed Number Represented by a Picture	• 4 Learning Palettes • 4 Learning Palette Card 44	Ask each child to match the problems and answers on Learning Palette Card 44.

Advanced Learners

Materials: pattern blocks (2 hexagons and 2 parallelograms)

- Place the pattern blocks in a row in front of the child; place the 2 parallelograms forming $\frac{2}{3}$ of a hexagon.
- Ask the child what mixed number the pattern blocks show; ask what fraction is missing in the partial hexagon. ($2\frac{2}{3}$, $\frac{1}{3}$)

• Representing and Writing Mixed Numbers (Part 2)

English Learners

- Draw and shade $2\frac{2}{3}$ circles; draw and shade $3\frac{3}{4}$ squares.
- Review **mixed number:** a number that contains both a whole number and a fraction
- Remind the child that a mixed number is a whole number with a fraction; remind him or her that the shaded drawings represent a mixture of whole shapes plus a fraction of the same shape.
- Ask the child to tell you the two mixed numbers pictured on the board.
- Ask the child to use a complete sentence to explain *mixed number.*

Spanish Cognate: mixed number número mixto

Inclusion Students

materials: 12 index cards: on six separate index cards show pictorially: $1\frac{1}{4}$, $1\frac{1}{3}$, $2\frac{4}{8}$, $2\frac{3}{4}$, $3\frac{1}{2}$, $3\frac{2}{3}$, using various shapes such as circles, squares, rectangles, or triangles; write the mixed numbers on the remaining six cards

- Ask the child to match each mixed number on a card to its picture.

At-Risk Students

- *materials:* pattern blocks
- On the board write: $4\frac{1}{2}$, $1\frac{1}{3}$, $3\frac{2}{3}$, $2\frac{1}{6}$.
- Ask the child to use pattern blocks (of his or her choosing) to show: $4\frac{1}{2}$, $1\frac{1}{3}$, $3\frac{2}{3}$, $2\frac{1}{6}$.
- Use the pattern blocks indicated to show the following values and ask the child to tell you the mixed numbers: (YG) $3\frac{5}{6}$, (RG) $1\frac{1}{3}$, (RB) $4\frac{2}{3}$, (BG) $3\frac{1}{2}$.

Advanced Learners

- On a sheet of paper write the following problem: Five pizzas were delivered for the party. Three and three-quarters of the pizzas were eaten. Draw a picture to show the five pizzas and shade them to show how much was eaten. Write mixed numbers to show how much pizza was eaten and how much remained.
- If necessary, remind the child that a quarter is equal to one fourth.

• Creating and Reading a Bar Graph With a Scale of 2

English Learners

- Write the following numbers on the board: 6, 7, 9, 6, 6, 4, 8, 7, 6, 4, 6, 2, 7, 6
- **mode:** the number that occurs most often in a list
- Explain that the word *mode* has several meanings in English; the meaning in mathematics is the number that occurs most often in a set of numbers.
- Point out that in the list of numbers on the board, several numbers appear more than once, but one number, which is the mode, appears most often; ask the child to identify this number. (6)
- Have the child use a complete sentence to explain *mode.*

Spanish Cognate: mode moda

Inclusion Students

materials: 15 nickels and a graph: on a sheet of paper create a simple 3-column and 8-row graph similar to that used during the lesson; label the base of the three columns *Before 1991*, *1991–1999*, and *2000–Now*; label the rows by 2's from 0 (at the base) to 16; and place boxes for tally marks below the column labels.

- Ask the child to look at the mint date of each coin and place a tally mark in the correct box.
- Ask him or her to count the tally marks for each number and to write the count under the tally marks box.
- Review that each box on the graph is equal to 2 and assist the child as needed in shading the graph with the values for each column; ask which column shows the most nickels minted.

At-Risk Students

materials: 20 nickels and a graph: on a sheet of paper create a simple 3-column and 10-row graph similar to that used during the lesson; label the base of the three columns *Before 1991*, *1991–1999*, and *2000–Now*; label the rows by 2's from 0 (at the base) to 20; and place boxes for tally marks below the column labels.

- Ask the child to look at the mint date of each coin and place a tally mark in the correct box.
- Ask him or her to count the tally marks for each number and to write the count under the tally marks box.
- Review that each box on the graph is equal to 2 and ask the child to shade the graph with the values for each column; ask which column shows the most nickels minted and review *mode.*

Advanced Learners

- Extension and Enrichment problem for Guided Class Practice 113, problem 3 (Extension and Enrichment Student Problem [page 27] at www.SaxonMathResources.com)

Extension and Enrichment Problem	Answer
Draw a square with a side of 6 cm. What is the perimeter? Draw a line of symmetry to divide the square into 2 rectangles. What is the perimeter of one of those rectangles?	a square with 6 cm sides; 24 cm; show a line of symmetry; 18 cm

- **Identifying Right Angles**

English Learners

- On the board draw a right angle; include the square angle marker.
- **right angle:** an angle that measures 90°; square corner
- **right triangle:** a triangle with a right angle
- Explain that the word *right* has been used previously to mean a direction, the opposite of left, and to mean correct, such as a right answer.
- Explain that when used to describe the angle of a figure, a right angle means a square corner; point to the angle drawn on the board and show that the small square indicates that it's a right angle.
- Draw a square and a rectangle on the board; in one add the square angle marker in one angle.
- Show that both of these figures have four right angles; note that the square angle marker is not always shown.
- Draw a right triangle without the angle marker; explain that a triangle that has a right angle is called a right triangle.
- Erase all the figures.
- Ask the child to draw a right triangle and to use a complete sentence to describe a *right angle.*

Inclusion Students

materials: geoboard and geobands

- Create a right triangle on the geoboard; review the terms *right angle* and *right triangle* with the child.
- Create an acute triangle on the geoboard; ask him or her if this figure has a right angle.
- Create other geometric figures of various sizes and shapes, with and without right angles, each time asking the child whether the figure has a right angle (and if so, how many); when the figure is a triangle, ask the child if it is a right triangle.

At-Risk Students

materials: 2 geoboards and geobands

- Create a right triangle on one geoboard and ask the child to copy the figure on his or her geoboard; review the terms *right angle* and *right triangle.*
- Create an acute triangle on one geoboard and ask the child to copy it; ask him or her if this figure has a right angle.
- Create other geometric figures of various sizes and shapes, with and without right angles, each time asking the child to copy the figure.
- Ask him or her whether the figure has a right angle (and if so, how many); when the figure is a triangle, ask the child if it is a right triangle.

Advanced Learners

- Learning Palette Enrichment Card E20, Using Geometric Terms

- **Multiplying by 2 Facts**

English Learners

- On the board write *factor* × *factor* = *product* and 10 × 9 = 90.
- **factors:** the numbers that are multiplied in a multiplication problem
- **product:** the answer in a multiplication problem
- Explain that when writing a multiplication problem, the factors are multiplied to find the product; if desired, compare these terms to the addends and sum in an addition problem.
- Ask the child to use a complete sentence to explain the terms *factors* and *product;* he or she may use the example on the board to help with the explanation.

Spanish Cognates: factors factores
product producto

Inclusion Students

materials: Teacher Fact Cards (Set S, Multiplying by 2 Facts, plus 3 × 2, 4 × 2, 5 × 2)

- Show the Teacher Fact Cards in random order; ask the child to say the answers.

At-Risk Students

- Math Center Activity 96

Objectives	Materials for Four Children	Teacher Directions
• Showing Repeated Addition as Multiplication	• 4 Learning Palettes • 4 Learning Palette Card 45	Ask each child to match the problems and answers on Learning Palette Card 45.

Advanced Learners

- Review the commutative property of multiplication.
- Ask the child to draw 4 rows of 5 circles and 5 rows of 4 circles on a sheet of paper; ask him or her to explain how the number of circles in the 2 groups shows the commutative property of multiplication.

Guide to Differentiated Instruction

- **Estimating Area**
- **Finding Area Using One-Inch Color Tiles**

English Learners

- Review **estimate:** to make a good guess about value or measurement based on given information
- Review why it is useful to be able to estimate such things as amount, length, weight, area, and so on. (There may not be the time or the tools to make an exact measurement, recognizing when an answer is not close to an estimate may show a calculation mistake, etc.)
- Ask the child to use a complete sentence to explain *estimate.*

Spanish Cognate: estimate estimar

Inclusion Students

materials: container of color tiles and four paper rectangles cut in the following dimensions: 2" × 5", 3" × 4", 3" × 5", 4" × 4"

- Give the first rectangle to the child and ask how many color tiles he or she thinks will be needed to cover the surface; have him or her write the estimate on a piece of paper; then ask the child to use the color tiles to find the area of the rectangle; ask him or her to find the difference between the estimated number of tiles and the actual number of tiles needed to cover the rectangle.
- Repeat with the remaining rectangles.

At-Risk Students

- Math Center Activity 98

Objectives	Materials for Four Children	Teacher Directions
• Finding Area by Counting Square Units	• 4 Learning Palettes • 4 Learning Palette Card 46	Ask each child to match the problems and answers on Learning Palette Card 46.

Advanced Learners

materials: 10" × 7" rectangle from a sheet of colored construction paper, ruler, and color tiles

- Ask the child to draw a 2" × 5" rectangle on a piece of white paper; ask him or her to show how many one-inch color tiles are needed to cover this area. (10)
- Assist the child in carefully cutting out the white rectangle; place the colored rectangle in front of the child so that the 7" side is across the top.
- Ask the child to find how many times the white rectangle will fit on the construction paper rectangle; have him or her start by placing the white rectangle in the upper left corner, with the 5" side across the top; help him or her trace the outline of the rectangle and then shift it down.
- The small rectangle will fit five times down the left side of the colored paper and twice (on end) on the right side.
- Explain that since the white rectangle is equal to 10 tiles and it fits 7 times on the paper, the multiplication number sentence 10 × 7 = 70 shows how many tiles will fit on the large rectangle.
- If desired, ask the child to find the lengths of the two sides of the large rectangle. (10" × 7")

• Acting Out and Drawing Pictures for "Equal-Groups" Stories

English Learners

- **symbol:** a mark that stands for something; a short way to write something (See Guided Class Practice 116A, #6)
- Remind the child that in mathematics and in everyday life he or she will see symbols instead of words that stand for something; an example is an arrow, which means *this way.*
- Explain that when they are understood, symbols are a quick and easy way to convey information; point out that sometimes symbols are called *signs*.
- Ask the child to identify each of the following symbols as you write them on the board: ¢, <, +, $, =, –, °, >, ×, ", and ☐ (as a right angle marker)
- If desired, explain the difference between a symbol and an abbreviation, which uses a letter or part of the word to mean something.
- Ask the child to use a complete sentence to explain *symbol.*

Spanish Cognate: symbol símbolo

Inclusion Students

materials: 4 strips of paper with the problems shown below, 16 color tiles, and 4 hundred-flats

- Prepare the following problems on strips of paper; read each problem in turn to the child and have him or her show the equal groups and say the total.
  1. There are 2 nests. Use hundred-flats to show the nests. There are 4 eggs in each nest. Use color tiles to show the eggs. How many eggs are there altogether?
  2. There are 3 basketball teams. Use hundred-flats to show the teams. There are 5 players on a basketball team. Use color tiles to show this. How many basketball players are there in all?
  3. There are 4 horses. Use hundred-flats to show the horses. There are 4 hooves on a horse. How many hooves are there in all? (Explain, if necessary, that hoof means foot.)
  4. There are 2 ants. Each ant has 6 legs. How many ant legs are there altogether?

At-Risk Students

- Ask the child to write two equal-groups stories and draw pictures to illustrate his or her stories. (If the child has trouble thinking of a subject, suggest writing about strings on guitars, bananas in bunches, or dimes in pockets.)

Advanced Learners

- Write the following problem for the child: Martin Lake has two areas where people are allowed to fish. Last Saturday, there were five people fishing in each of these areas. Each of the people fishing caught four fish. How many fish were caught at Martin Lake last Saturday? (40 fish)
- Ask the child to draw a picture to help him or her find the answer.
- If the child has trouble solving the problem, suggest that he or she should first find the number of people who were fishing.

- **Writing Number Sentences for "Equal-Groups" Story Problems**

English Learners

- On the board write 4 + 9 = 13 and 2 × 6 = 12.
- **total:** the whole number or amount; complete
- Explain that *total* is a general term meaning the whole amount, and it can be used to describe the answer in either an addition or a multiplication problem.
- Point out that 13 is the total or sum when adding 4 and 9, and that 12 is the total or product when multiplying 2 by 6.
- Ask the child to use a complete sentence to explain *total.*

Spanish Cognate: total total

Inclusion Students

materials: 10 index cards: 5 of the cards showing the problems pictorially (with simple drawings) and 5 of the cards showing the number sentence: 3 spiders with 8 legs each (3 × 8 = 24); 3 shirts with 4 buttons each (3 × 4 = 12); 5 squares with 3 stars inside each (5 × 3 = 15); 3 cookies with 6 chips each (3 × 6 = 18); and 4 vases with 4 flowers each (4 × 4 = 16)

- Ask the child to match the number sentence to the equal groups picture; explain and correct any errors only when the child has completed the activity.

At-Risk Students

- Math Center Activity 100

Objectives	Materials for Four Children	Teacher Directions
• Identifying a Number Sentence for an Equal Groups Picture	• 4 Learning Palettes • 4 Learning Palette Card 47	Ask each child to match the problems and answers on Learning Palette Card 47

Advanced Learners

- Extend and Challenge CD, Activity 9, At the Amusement Park
- After the child finishes the activity, use the question(s) on the recording form to evaluate his or her understanding. (Monitoring Student Progress binder, Extend and Challenge CD booklet)

Guide to Differentiated Instruction

- **Identifying Intersecting Lines**
- **Identifying Perpendicular Lines**

English Learners

materials: Math Folder (Line Segments)

- On the board draw a pair of intersecting line segments (non-perpendicular) and a pair of perpendicular line segments (showing the right angle mark)
- **intersect:** to cross
- **perpendicular:** lines that intersect to form right angles
- Explain that when line segments meet, they intersect, and that the place where they meet or cross is called the *intersection.*
- Explain that when line segments meet or cross to form a right angle, the lines are called *perpendicular;* ask the child to show you the perpendicular line segments in the Math Folder and to explain the difference between parallel lines and those that *intersect.*
- Ask the child to use a complete sentence to explain *perpendicular.*

Spanish Cognates:	intersect	intersecar
	perpendicular	perpendicular

Inclusion Students

materials: geoboard and geobands

- Remind the child that lines that intersect at right angles (square corners) are perpendicular.
- On the geoboard create a sequence of pairs of line segments; have the pairs with various lengths and orientations; have some parallel, some perpendicular (including an *L*), and some intersecting but not perpendicular.
- Ask the child to identify each pair as parallel, intersecting, or perpendicular; if he or she says that a perpendicular pair is intersecting, explain that it is correct, but ask what else they are.

At-Risk Students

- Math Center Activity 101

Objectives	Materials for Four Children	Teacher Directions
• Identifying Geometric Terms	• 4 Learning Palettes • 4 Learning Palette Card 48	Ask each child to match the problems and answers on Learning Palette Card 48.

Advanced Learners

- Extension and Enrichment problem for Guided Class Practice 118, problem 1 (Extension and Enrichment Student Problem [page 28] at www.SaxonMathResources.com)

Extension and Enrichment Problem	Answer
Crayons are sold in boxes of 8. Mrs. Martinez bought 3 boxes and then gave 1 box away. How many crayons does she have left?	16 crayons

- **Subtracting Three-Digit Numbers and Money Amounts**
- **Estimating Differences**

English Learners

materials: container with 20 color tiles

- **reasonable:** based on good sense (sensible); explainable; thought out
- Explain that when people make decisions based on available information, they are using reasoning; ask the child if he or she would wear a bathing suit in cold weather and to explain why or why not.
- Continue that in the same way, when people make estimates based on available information, they use reasoning; they can judge whether something makes sense or not and when something makes sense, it is reasonable.
- Ask which is a better estimate of the number of color tiles in the container, 20 or 200, and ask why it is a reasonable estimate.
- Ask the child to use a complete sentence to explain the word *reasonable.*

Spanish Cognate: reasonable razonable

Inclusion Students

The child will practice subtracting three-digit numbers and money amounts.

- Write *647 pennies* and *225 pennies* on the board; say that you started with 647 pennies and gave 225 to a friend; ask the child to describe how to write the problem.
- Ask the child to solve the subtraction, starting with the ones' place.
- Ask the child to write these money amounts with the dollar sign and decimal points, keeping the columns aligned; ask him or her to repeat the subtraction, being sure to write the dollar sign and decimal point in the answer.
- Repeat with 562 pennies and 318 pennies and with 151 pennies and 116 pennies.

At-Risk Students

The child will practice estimating sums and differences.

- Review the concept of rounding numbers; explain that just as numbers can be rounded to the nearest 10, they can be rounded to the nearest 100.
- Remind the child that rounding numbers makes mental estimation much easier and more accurate; explain that mental estimation is a skill that is useful since it can be used to decide whether math answers are reasonable, it can be used when paper and pencil are not available, and so on.
- Ask the child to estimate the answer by rounding the numbers to the nearest 100, and then find the answer to the following problems: $604 + 495 =$, $98 + 113 =$, $312 + 280 =$, $215 - 95 =$, $391 - 206 =$.

Advanced Learners

- Learning Palette Enrichment Card E21, Subtracting Three-Digit Numbers
- Learning Palette Enrichment Card E22, Subtracting and Adding Money Amounts

Guide to Differentiated Instruction Lesson 120-1

- **Dividing a Set of Objects Into Equal Groups**
- **Multiplying by 3 Facts**
- **Making a Table to Solve a Problem**

English Learners

- Cut a large capital *Z* from a piece of paper.
- Review **flip:** to turn over a figure from front to back or back to front (See Meeting 120-1 Today's Pattern.)
- Explain that to change the figure Z, to a backward Z, you flip the figure—turn it over from front to back; demonstrate this with the paper letter.
- Show the same thing with the next pair of letters, backward N and N; demonstrate this with the paper letter and use the word flip as you explain.
- Ask the child to use a complete sentence to explain *flip;* he or she may use the paper letter as an example.

Inclusion Students

The child will practice multiplying by 3.

materials: Teacher Fact Cards (Set T, Multiplying by 3 Facts, plus 2 × 3, 4 × 3, 5 × 3)

- Show the Teacher Fact Cards in random order; ask the child to say the answers.

At-Risk Students

materials: Problem-Solving Worksheet 120A and Performance Task Worksheet 120B

- Review with the child the problem-solving strategies and Problem-Solving Worksheet 120A and Performance Task Worksheet 120B.

Advanced Learners

- Extend and Challenge CD, Activity 10, Looking for Patterns
- Have the child complete Extend and Challenge CD Activity 10 Worksheet.
- After the child finishes the activity, use the question(s) on the recording form to evaluate his or her understanding. (Monitoring Student Progress binder, Extend and Challenge CD booklet)

Guide to Differentiated Instruction — Lesson 120-2

- **Describing the Likelihood of an Event**
- **Predicting the Outcome of a Probability Experiment**

English Learners

materials: 4 yellow color tiles and a paper bag

- Draw a line segment on the board; at the left endpoint write *impossible* and at the right endpoint write *certain.*
- **certain:** absolutely sure to happen
- **impossible:** cannot happen
- Review that when there is the possibility or chance of something happening, it can be seen as a line from where it has no chance at all of happening (impossible) to the opposite end where it is absolutely sure to happen (certain).
- Show the child the four yellow tiles as you place them in the bag; ask him or her whether it is certain or impossible that a tile taken from the bag will be yellow; ask why; ask whether it is certain or impossible that a tile taken from the bag will be red; ask why.
- Have the child use a complete sentence to explain *certain* and *impossible.*

Spanish Cognate: impossible imposible

Inclusion Students

materials: die

- Review that likelihood is the possibility or chance that something will happen; it can go from something that has no chance of happening (impossible) to something that is sure to happen (certain); explain that between impossibility and certainty, things may happen with more or less likelihood.
- Show the child the die and ask him or her how likely it is to roll a number less than 7; … to roll a 9; … to roll a number greater than 0; … to roll the letter *b*.
- Explain that rolling a 1, 2, 3, 4, 5, or 6 are all equally likely, and there is no way to predict which of these numbers will come up.

At-Risk Students

- Math Center Activity 103

Objectives	Materials for Four Children	Teacher Directions
• Identifying the Likelihood of an Event	• 4 Learning Palettes • 4 Learning Palette Card 50	Ask each child to match the problems and answers on Learning Palette Card 50.

Advanced Learners

- Learning Palette Enrichment Card E23, Identifying the Likelihood of an Event

- **Making and Labeling an Array**

English Learners

materials: 12 color tiles

- **array:** an orderly arrangement of objects in the shape of a rectangle
- **row:** a horizontal arrangement of items
- Review **column:** a vertical arrangement of items
- Describe an array, rows, and columns while arranging the color tiles in a 3 × 4 array.
- Write 3 × 4 on the board, pointing out that the first number indicates the number of rows and the second number indicates the number of columns.
- Write 4 × 3 on the board and rearrange the tiles to reflect this.
- Write 6 × 2 on the board and ask the child to create a 6 × 2 array.
- Ask the child to tell you what the words *row* and *column* mean, and to use them in a sentence to explain the word *array.*

Inclusion Students

materials: 12 color tiles and 4 index cards with one of the following on each: 2 × 3; 4 × 2; 3 × 4; 5 × 2

- Ask the child to use the color tiles to create each array.
- Remind the child that the first number indicates the number of rows and the second number indicates the number of columns; review *row* and *column* if needed.

At-Risk Students

materials: 16 color tiles and 6 index cards with one of the following on each: 2 × 3; 4 × 2; 3 × 4; 4 × 3; 4 × 4; 5 × 2

- Ask the child to use the color tiles to create each array.
- Remind the child that the first number indicates the number of rows and the second number indicates the number of columns.
- Create a 2 × 6 array and ask the child to name it; repeat with a 3 × 2 array.

Advanced Learners

materials: 24 color tiles

- Ask the child to find and record as many arrays as possible using all 24 color tiles. (1 × 24, 2 × 12, 3 × 8, 4 × 6, 6 × 4, 8 × 3, 12 × 2, 24 × 1)
- Explain that an array may have only one row or column.

- **Writing Number Sentences for Arrays**

English Learners

- Using X's, draw a 5 × 3 array and a 4 × 5 array on the board; write 5 × 3 = 15.
- Review **array:** an orderly arrangement of objects in the shape of a rectangle
- Review that when writing the numbers in an array, the first shows the number of rows and the second shows the number of columns.
- Point out that the completed number sentence shows the number of X's in the first array.
- Ask the child to write a number sentence for the second array and explain his or her work.

Inclusion Students

materials: 6 index cards with one of the following number sentences on each: 2 × 3 = 6, 4 × 2 = 8, 3 × 4 = 12, 4 × 3 = 12, 5 × 2 = 10, 5 × 3 = 15; and 6 index picture cards showing the same arrays made with X's or O's

- Ask the child to match each array to the correct picture card.
- Review the meanings of row and column, if needed, and review which number is which when writing an array.

At-Risk Students

- Math Center Activity 104

Objectives	Materials for Four Children	Teacher Directions
• Making an Array for a Number Sentence • Identifying a Number Sentence for an Array	• 8 index cards *Write one of the following on each of the index cards: 2 × 3; 3 × 2; 2 × 4; 4 × 2; 3 × 4; 4 × 3; 2 × 5; 5 × 2; On the back of each card draw the array and label it.* • basket of color tiles • Center 104 Master *Make 4 copies.*	Ask each child to choose a card and use color tiles to make the array on Center 104 Master. When the child finishes, ask the child to turn the card over and check the array. Repeat with three more cards.

Advanced Learners

- Extension and Enrichment problem for Guided Class Practice 122, problem 5 (Extension and Enrichment Student Problem [page 28] at www.SaxonMathResources.com)

Extension and Enrichment Problem	Answer
I have 2 quarters, 2 dimes, 2 nickels, and 4 pennies. How much money do I have? If I add 1 dozen pennies to my total, how much money will I have altogether? Write the new money amount in 2 different ways.	84¢; \$0.96 and 96¢

• Telling and Showing Time to the Quarter Hour

English Learners

materials: demonstration clock

- On the board draw a circle divided into fourths with a horizontal and a vertical line.
- Review **quarter:** one fourth
- **quarter hour:** a measurement of time equal to 15 minutes; one fourth of an hour
- Review that a quarter means a fourth, just as a quarter coin is one fourth of a dollar.
- Point to the circle divided into fourths and show that it is like a clockface divided into quarter hours or fourths of an hour.
- Using the demonstration clock, show various times like 7:45, 3:15, 2:30, and so on, saying them both as digital times and using the wording *quarter to, quarter of, quarter after, quarter past,* and *half past;* show additional times and ask the child to say them both ways.
- Have the child use a complete sentence to explain *quarter hour.*

Spanish Cognates:

quarter	cuarto
quarter hour	cuarto de hora

Inclusion Students

materials: demonstration clock and student clock

- Write the following times on the board and ask the child to say them using both digital time and the wording *quarter to, quarter of, quarter after, quarter past,* and *half past:* 6:45, 1:15, 4:30, 9:15, 2:45.
- Repeat the activity showing the following times on the demonstration clock: 7:15, 3:45, 12:30, 5:45, 2:30.
- Slowly say the following times (using quarter after, etc.) and ask the child to show them on the student clock: 1:15, 1:30, 10:45.

At-Risk Students

- Math Center Activity 105

Objectives	Materials for Four Children	Teacher Directions
• Telling Time to the Quarter Hour	• 4 Learning Palettes • 4 Learning Palette Card 51	Ask each child to match the problems and answers on Learning Palette Card 51.

Advanced Learners

- Extension and Enrichment problem for Guided Class Practice 123, problem 2 (Extension and Enrichment Student Problem [page 29] at www.SaxonMathResources.com)

Extension and Enrichment Problem	Answer
The piano recital began at quarter past 3 in the afternoon. It lasted one and one half hours. What time was the piano recital over?	4:45 p.m.

- **Identifying and Showing Transformations: Translations, Rotations, and Reflections**

English Learners

materials: playing card

- **turn:** to rotate a shape around a fixed point
- **slide:** to transform a shape by moving it from one place to another on a flat surface without turning it
- Review **flip:** to turn over a figure from front to back and back to front
- Explain that mathematicians use different words to describe the movement of things, and that often objects move in more than one of these ways at the same time.
- Turn, slide, and flip the playing card as you describe and explain each individual movement; then combine two of the movements as you move the card.
- Have the child use a complete sentence to describe the difference between the movements *flip, slide,* and *turn;* allow him or her to use the card to demonstrate.

Inclusion Students

materials: doll, action figure, or card

- Review the words *flip, turn,* and *slide* with the child.
- Demonstrate and describe each separate movement with the doll, action figure, or card.
- Ask the child to repeat the movement with the doll, action figure, or card as you say each word.
- Combine movements, such as slide and flip, and ask the child to say what movements were used.

At-Risk Students

materials: 2 playing cards

- Demonstrate and describe each separate movement (flip, turn, and slide) with the card.
- After showing all three movements, ask the child to repeat the movements with the card as you say each word.
- Using the card, combine movements, such as slide and flip, and ask the child to say what movements were used.
- Place the cards in different orientations and ask how to cover one card with the other.

Advanced Learners

materials: globe (if available, or a sphere) and a flashlight

- Explain that the Earth is a sphere and that it turns around a line that runs from the top of the Earth (the North Pole) through the middle of the sphere to the bottom of the Earth (the South Pole).
- Explain that points on the Earth completely turn, or *rotate,* around this line once each day.
- Ask the child to point the flashlight at the globe (or at the sphere in your hand); turn the globe slowly; explain that the flashlight represents the light of the sun, and that the part of the globe facing the sun is in daylight and the part facing away shows night.

- **Dividing a Set of Objects Into Equal Groups**
- **Multiplying by 4 Facts**

English Learners

materials: a deck of playing cards

- **deck:** a pack or set of cards
- **playing cards:** rectangular cardboard pieces with symbols on one side
- Explain that the word *deck* has several meanings in English, such as the floor of a ship or a wooden structure like a porch; when used with playing cards it means a complete pack of cards.
- Explain that playing cards, such as those used for activities in class, are commonly used for many kinds of games.
- Show the child the symbols for the four suits and show that each suit has an ace, numbered cards two through ten, and the face cards jack, queen, and king.
- Have the child use a complete sentence to describe a *deck of cards.*

Inclusion Students

The child will practice multiplying by 4.

materials: Teacher Fact Cards (Set U, Multiplying by 4, plus 2 × 4, 3 × 4, 5 × 4)

- Show the Teacher Fact Cards in random order; ask the child to say the answers.

At-Risk Students

- Math Center Activity 106

Objectives	Materials for Four Children	Teacher Directions
• Identifying Paired Numbers in Real-Life Situations	• 4 Learning Palettes • 4 Learning Palette Card 52	Ask each child to match the problems and answers on Learning Palette Card 52.

Advanced Learners

- Learning Palette Enrichment Card E24, Identifying the Missing Numbers in a Counting Sequence

Guide to Differentiated Instruction

- **Choosing a Survey Question and Choices**
- **Representing Data Using a Graph**

English Learners

- On the board write: What is your favorite school subject?
- **survey:** to collect information by asking a group of people questions that have answer choices
- Explain that the word *survey* is used for both the action of collecting information and also for the information that is collected.
- Explain that people take surveys for different reasons; for example, makers of food products take surveys to find out what people like best and are most likely to buy.
- Point out that most surveys provide a limited number of choices for answers; this makes collecting and graphing the information easier.
- Ask the child to name four possible answer choices for the question on the board; write these on the board below the question.
- Ask him or her to use a complete sentence to explain *survey.*

Inclusion Students

- On the board write: What is your favorite zoo animal?
- Ask the child to name five possible answer choices for the question on the board; write these on the board below the question.
- Say: Let's pretend we have asked a group of children this question and tallied their answers.
- After each animal write one of the following numbers using tally marks: 5, 2, 6, 4, 7.
- Ask the child to tell you each number shown by the tally marks.
- Ask the child various questions like: Which zoo animal is the favorite of most children? and How many more children like [animal A] than like [animal B]?

At-Risk Students

- Math Center Activity 107

Objectives	Materials for Four Children	Teacher Directions
• Reading a Bar Graph With a Scale of 2 • Reading a Tally Chart	• 4 Learning Palettes • 4 Learning Palette Card 53	Ask each child to match the problems and answers on Learning Palette Card 53.

Advanced Learners

Journal Writing (Overview 13)

- Ask the child if he or she were going to take a survey, what the question would be.
- Have the child write the question and describe how he or she would collect the information.

- **Locating and Graphing Points on a Coordinate Graph**

English Learners

- On the board draw a 5 × 5 coordinate graph like that used in the lesson: a dot array with labels 0–4 along both axes.
- **coordinate graph:** a diagram containing intersecting perpendicular lines on which points are identified by pairs of numbers
- **coordinates:** the numbers that show the location of a point on a coordinate graph
- **origin:** the point of intersection of the horizontal and vertical number lines on a coordinate plane
- Explain that a coordinate graph may show vertical and horizontal lines which meet where the dots are shown, but for now the dots are enough to show how the graph works.
- Show the origin and explain that it is point (0, 0); explain that just as the word origin often means the beginning of something, the origin of the graph is the base point that other points are counted from.
- Draw a star at point (2, 4) and a square at point (4, 3); explain that these points can be located by pairs of numbers called coordinates.
- Explain that the first number shows the horizontal distance from the origin and the second number is the vertical distance.
- Ask the child to use a complete sentence to explain the terms *origin* and *coordinates;* he or she may use the graph to help with the explanation.

Spanish Cognates:	coordinate graph	gráfica con coordenades
	coordinates	coordenadas
	origin	origen

Inclusion Students

- Draw a 5 × 5 coordinate graph on a sheet of paper, using a dot array with labels 0–4 along both axes.
- Review the meanings of the terms origin and coordinates; explain how coordinates are named.
- Name several pairs of coordinates and ask the child to mark the points; assist as needed.

At-Risk Students

- Math Center Activity 108

Objectives	Materials for Four Children	Teacher Directions
• Locating Points on a Coordinate Graph	• 4 Learning Palettes • 4 Learning Palette Card 54	Ask each child to match the problems and answers on Learning Palette Card 54.

Advanced Learners

- Learning Palette Enrichment Card E27, Identifying Ordered Pairs on a Coordinate Graph

- **Showing and Counting Back Change for $1.00**

English Learners

materials: ruler marked with 40¢ price tag, 1 one-dollar bill, 2 quarters, and 5 dimes

- **change:** the money you get back after paying for something
- Explain that the word *change* has several meanings in English; the child is probably familiar with the meaning of becoming different or using something different.
- Review that when buying something at the store, the shopper does not always have the exact amount of money needed; when this happens, the shopper will give the clerk a larger amount of money, such as rounding up to a dollar, and receive change in return.
- Explain that *change* is the difference between the cost of an item and the amount given to the clerk; it is the extra amount of money that is returned.
- Use the coins to show that 40¢ + 60¢ = $1.00; demonstrate that if the child were to buy the ruler with the dollar bill, the change would be 60¢, since $1.00 – 40¢ = 60¢.
- Ask the child to use a complete sentence to explain *change*.

Inclusion Students

materials: 3 classroom items with price tags (from Lesson 127), 3 quarters, 5 dimes, and 5 nickels

- Ask the child to choose one item and to show the amount he or she would receive in change if he or she paid for the item with a one-dollar bill.
- Review that the cost of the item plus the amount of the change will equal one dollar; if desired, the child may show one dollar in coins and then subtract the cost of the item to find the change.
- Review the process of *counting back* starting with the cost of the item.
- Repeat with the other two items.

At-Risk Students

- Math Center Activity 109

Objectives	Materials for Four Children	Teacher Directions
• Making and Counting Change for $1.00	• 8 items with price tags (from Lesson 127) • 2 cups with 3 quarters, 5 dimes, and 5 nickels	Pair the children. Ask each child to choose an item and show the change he/she will receive if he/she pays for the item with a one-dollar bill. Ask the children to take turns counting back the change to their partner. Ask the children to repeat the activity with different items.

Advanced Learners

- Learning Palette Enrichment Card E25, Determining Change

- **Dividing by 2**

English Learners

materials: Workmat 28 and 14 color tiles

- **division:** an operation to separate a number of objects into equal groups
- Remind the child that the action word *divide* means to separate into equal groups.
- Remind the child that in previous lessons, he or she separated even or odd numbers of tiles into equal groups; explain that this is the process of division.
- Ask the child to use the 2 circles on the workmat to divide the tiles into 2 equal groups.
- Have the child use a complete sentence to explain *division.*

Spanish Cognate: division división

Inclusion Students

materials: Workmat 28 and 16 color tiles

- Remind the child that in previous lessons, he or she divided even or odd numbers of tiles in half, which is exactly the same as dividing by 2.
- Remind him or her that when an odd number of tiles is divided by 2, one tile, called the remainder, will be left; use tiles to demonstrate dividing 5 by 2 and write $5 \div 2 = 2$ R1 on the board.
- On the board write $16 \div 2 =$; ask the child to show this division with color tiles on the workmat; ask him or her how many tiles are in each equal group and if there was a tile left over.
- Write the solution on the board.
- Repeat with: $11 \div 2 =$; $8 \div 2 =$; $7 \div 2 =$.

At-Risk Students

materials: Workmat 28 and 20 color tiles

- On the board draw a rectangle divided in half with a vertical line; inside each half draw 3 circles; write ½ above each section.
- Point to the drawing while explaining that dividing a group of objects *by 2,* or *in 2,* is the same as dividing the group in half, since half of the group is one of two equal parts of the whole group.
- Remind the child that in previous lessons, he or she divided even or odd numbers of tiles in half; remind him or her that when an odd number of tiles was divided by 2, one tile, called the remainder, was left.
- On the board write $20 \div 2 =$; ask the child to show this division with color tiles on the workmat; ask how many tiles are in each equal group and if there is a remainder.
- Write the solution on the board.
- Repeat with: $16 \div 2 =$; $15 \div 2 =$; $8 \div 2 =$; $11 \div 2 =$.

Advanced Learners

- Learning Palette Enrichment Card E26, Completing a Function Table

Guide to Differentiated Instruction

- **Finding the Area of a Rectangle**

English Learners

materials: 16 color tiles and a 4" × 4" square cut from a sheet of paper

- Review **area:** the number of square units needed to cover a surface
- Review that area is the measure of the size of a flat surface; the surface may be small, such as a paper shape or the top of a table, or much larger, such as a piece of land.
- Ask the child to cover the 4" × 4" square with color tiles and to count the number of tiles needed; review that the result, 16 color tiles, is the area of the square.
- Ask the child to use a complete sentence to explain *area.*

Spanish Cognate: area área

Inclusion Students

materials: 18 color tiles; a 4" × 4" square, a 2" × 5" rectangle, and a 3" × 6" rectangle cut from a sheet of paper

- Ask the child to cover the 4" × 4" square with color tiles and to count the number of tiles needed; review that the result, 16 color tiles, is the area of the square.
- Repeat with the 2" × 5" and 3" × 6" rectangles.

At-Risk Students

materials: 18 color tiles; a 4" × 4" square, a 2" × 5" rectangle, and a 3" × 6" rectangle cut from a sheet of paper

- Ask the child to cover the 4" × 4" square with color tiles and to count the number of tiles needed; review that the result, 16 color tiles, is the area of the square.
- On the board write 4 × 4 = 16; on the square, point to the 4 tiles that make the top row and the 4 tiles that make the left column; explain that like an array, area can be shown with a multiplication number sentence.
- Repeat these steps with the 2" × 5" and 3" × 6" rectangles.

Advanced Learners

materials: a 12" × 12" (or 6" × 6") piece of paper, and 2" × 2" and 3" × 3" (or other size square) self-stick notes

- Review that the number of square units needed to cover an area depends on the size of the square unit being used to measure by.
- Ask the child to find the area of the piece of paper using the smaller squares; ask him or her to record the number.
- Ask the child now to find the area using the larger squares; ask him or her to record the number.
- Have the child write a paragraph comparing and explaining the results found when measuring with two different size square units.

- **Multiplying by 0 Facts**
- **Identifying the Multiples of 2, 3, 4, and 5**
- **Solving a Problem by Guessing and Checking**

English Learners

materials: the Meeting Board counting strips for reference

- On the board write the following number sentences in a column: $1 \times 3 = 3$, $2 \times 3 = 6$, $3 \times 3 = 9$, $4 \times 3 = 12$; above the column of products write *Multiples of 3:*.
- **multiples:** numbers that are the products of multiplying a number by 1, 2, 3, 4, and so on
- Point to the equations written on the board; show the column of counting numbers and that each is multiplied by the same number, 3; point to the answers to the multiplication problems and explain that they are the multiples of 3.
- Refer to the counting by 3's strip on the Meeting Board; say that when counting by 3's you are saying the multiples of 3, and have the child count by 3's to 30.
- Have the child use a complete sentence to describe *multiple.*

Spanish Cognate: multiple múltiple

Inclusion Students

The child will identify the multiples of 3 and 4.

materials: photocopies of Activity 10 Worksheet (Multiples of 3) and Activity 11 Worksheet (Multiples of 4) in the Extend and Challenge CD booklet (Monitoring Student Progress binder)

- Have the child show the respective multiples on the charts on the worksheets.
- If the child has difficulty with fine motor skills, he or she may find it hard to quickly and neatly color in the squares; in this case allow the child to draw an X over the numbers or to use a stamp pad and small stamp (or pencil eraser) to mark the numbers.
- Ask the child to show you any patterns he or she notices on the charts.

At-Risk Students

materials: Problem-Solving Worksheet 130A and Performance Task Worksheet 130B

- Review with the child the problem-solving strategies and Problem-Solving Worksheet 130A and Performance Task Worksheet 130B.

Advanced Learners

- Extension and Enrichment problem for Problem-Solving Worksheet 130A (Extension and Enrichment Student Problem [page 31] at www.SaxonMathResources.com)

Extension and Enrichment Problem	Answer
Tan has a domino. One half has twice as many dots as the other half. The sum of the dots is an odd number greater than 4. Show what the dots on his domino look like.	a domino with 3 dots on one half and 6 dots on the other half

• Making and Using a Multiplication Table

English Learners

materials: Lesson Worksheet 130-2

- **multiplication table:** a chart that shows the multiplication facts
- Review that a table is a way of showing information in rows and columns; the multiplication table created in the lesson shows multiplication facts covered so far (from 0 × 0 to 9 × 5 [5 × 9]).
- Review that a multiplication answer is found by finding one of the factors (numbers being multiplied) in the left column, the other factor in the top row, and running your fingers along the row and column until they meet.
- Explain that it does not matter which number from a multiplication problem is found in the left column; show the child that 3 × 5 and 5 × 3 both will result in the answer 15.
- Ask the child to find the answer to 2 × 7 and 3 × 4; ask him or her to use a complete sentence to explain a *multiplication table.*

Spanish Cognate: multiplication table tabla de multiplicación

Inclusion Students

materials: Lesson Worksheet 130-2 and sheet of lamination paper

- Review how to use a multiplication table; if this has not already been done, assist the child as he or she fills in the rest of the answers on Lesson Worksheet 130-2 (from 0 × 0 to 9 × 5 [5 × 9]).
- Review that each column shows the multiples for the number at the top of the column or that each row shows the multiples for the number at the left of the row.
- Explain that it does not matter which number from a multiplication problem is found in the left column; show the child that 3 × 5 and 5 × 3 both will result in the answer 15.
- Ask the child to find the answer to: 2 × 7, 3 × 4, 9 × 0, 5 × 4, 6 × 4, 8 × 2.
- Laminate the worksheet and ask the child to take it home for additional review.

At-Risk Students

- Math Center Activity 110

Objectives	Materials for Four Children	Teacher Directions
• Identifying Missing Numbers on a Multiplication Chart	• 4 Learning Palettes • 4 Learning Palette Card 55	Ask each child to match the problems and answers on Learning Palette Card 55.

Advanced Learners

- Learning Palette Enrichment Card E28, Completing a Function Table

- **Selecting the Appropriate Tool to Measure Mass**
- **Measuring Weight (Mass) Using Metric Units**

English Learners

materials: one dollar bill and 4 pennies

- **gram:** a small metric unit used to measure weight
- **mass:** the amount of matter in an object found by using a scale; weight
- Remind the child that in the United States, weight is generally measured using pounds, but in many other countries and in science, weight is generally measured using grams and kilograms (1000 grams).
- Review that a gram is a small unit of mass; ask the child to hold and feel the weight of the dollar bill, which weighs about 1 gram; ask him or her to hold and feel the weight of the 4 pennies, which weigh about 10 grams.

  Note: While mass and weight are not the same, at this level they can be used to mean the same thing.
- Ask the child to use a complete sentence to explain *gram* and *mass.*

Spanish Cognates: gram gramo
mass masa

Inclusion Students

materials: balance, container of paper clips, and 2 or 3 classroom objects weighing less than the container of paper clips

- Assist the child as he or she weighs each object; record the weight of each.

At-Risk Students

- Math Center Activity 111

Objectives	Materials for Four Children	Teacher Directions
• Measuring Mass Using Nonstandard Units • Ordering Objects from Lightest to Heaviest	• 5 classroom objects weighing less than the container of paper clips • balance • container of paper clips • Center 111 Master *Make 1 copy for each child in class.*	Pair the children. Ask each pair to weigh an object using paper clips, and have children write the name of the object and the number of paper clips used on Center 111 Master. Ask the children to repeat with the other four objects. Ask each pair to put the objects in order from lightest to heaviest.

Advanced Learners

Journal Writing (Overview 14)

- Have the child write one or two paragraphs about something that weighs more than he or she does and something that weighs less.

- **Doubling a Number**

English Learners

materials: 10 color tiles

- On the board write: 5 + 5 = 10, 2 × 5 = 10.
- Review **doubles:** two of the same number
- **twice:** two times
- Remind the child that early in the year, he or she learned addition doubles facts; this was the addition of two of the same number, like 5, to find their sum.
- Divide the color tiles into two groups of five, to show that five plus five is equal to ten.
- Explain that this same arrangement of tiles also represents 2 groups of 5 or 2 times 5.
- Explain that in both cases the meaning of doubling is the same: it is adding two of the same number together or multiplying a number by two; either way the answer is twice the original number.
- Ask the child to use a complete sentence to explain *doubles* and *twice.*

Spanish Cognate: doubles dobles

Inclusion Students

materials: 40 pennies

- On the board write: 5¢ + 5¢ = 10¢, 2 × 5¢ = 10¢.
- Use the pennies to make two groups of five, showing that five plus five is equal to ten.
- Explain that this same arrangement of pennies also represents 2 groups of 5 or 2 times 5.
- Explain that in both cases the meaning of doubling is the same: it is adding two of the same number together to find their total or multiplying that number by two.
- On the board write 20¢ and ask the child to double this amount; the child may use pennies, if necessary, to visualize the doubling process.
- Repeat with: 15¢, 12¢.

At-Risk Students

materials: five store coupons with different values from 10¢ to 50¢ (from the lesson)

- Remind the child, if necessary, how coupons are used and that some stores will double their value.
- Ask the child to double each coupon; write the original and doubled values on the board.

Advanced Learners

- Extension and Enrichment problem for Guided Class Practice 132, problem 1 (Extension and Enrichment Student Problem [page 31] at www.SaxonMathResources.com)

Extension and Enrichment Problem	Answer
Shelley has 3 dolls. Marybeth has twice as many dolls as Shelley has. If each of Marybeth's dolls is wearing a pair of shoes, how many doll shoes does Marybeth have?	12 doll shoes

- **Dividing a Set of Objects Into Equal Groups**

English Learners

materials: a playing card with diamonds, a tan pattern block, and a picture of a diamond ring

- **diamond:** colorless precious stone; a parallelogram with equal length sides (a rhombus) (Note: a diamond is one of the figures shown in the Student Glossary under the entry *polygon.*)
- Explain that there are several meanings for the word *diamond* in English; in Lesson 133, the children pretended that color tiles were the valuable stones used in jewelry; show the child the picture of the diamond ring.
- Remind the child that *diamonds* is one of the four suits in a deck of cards and in general the diamond shape is shown by the tan pattern block with an angle pointed down; the child may be familiar with a baseball diamond.
- Ask the child to use a complete sentence to explain at least one meaning of diamond.

Spanish Cognate: diamond diamante

Inclusion Students

materials: 16 color tiles

- Show that 12 color tiles may be divided into either 6 groups of 2 or 2 groups of 6; explain that any even number (over 2) may be divided into groups with two each or into two groups.
- Ask the child to find two other ways to divide 12 tiles into equal groups. (3 × 4, 4 × 3)
- Repeat with 16 color tiles. (8 × 2, 2 × 8, 4 × 4)

At-Risk Students

- Math Center Activity 113

Objectives	Materials for Four Children	Teacher Directions
• Finding the Answer to a Division Problem Using a Picture	• 4 Learning Palettes • 4 Learning Palette Card 56	Ask each child to match the problems and answers on Learning Palette Card 56.

Advanced Learners

- Extension and Enrichment problem for Guided Class Practice 133, problem 4 (Extension and Enrichment Student Problem [page 32] at www.SaxonMathResources.com)

Extension and Enrichment Problem	Answer
Sam, Ally, Tracee, and Joe will share 2 dozen doughnuts equally for breakfast. How many doughnuts will each person have? If two friends join them for breakfast and share the doughnuts equally, how many doughnuts would each have?	6 doughnuts; 4 doughnuts

- **Recording Information on a Graph**

English Learners

materials: pair of dice

- On the board write: probability experiment.
- **probability experiment:** a series of tests to find out how likely it is that something will happen
- Explain that the term is made up of two words; the class has discussed *probability,* which is the likelihood of something happening. The second word, *experiment,* means a scientific study; it is usually a set of tests to find some result.
- Explain that during an experiment an activity is completed many times, not just once.
- Roll the dice several times to demonstrate why it is necessary to run a series of tests—successive rolls will have different results; explain that it is the job of the person running an experiment to see all the possible results and to find how common each result is.
- Explain that 7 is the number most likely to be rolled; an experiment should show this.
- Ask the child to use a complete sentence to explain *probability experiment.*

Spanish Cognate: probability experiment experimento de la probabilidad

Inclusion Students

materials: 15 color tiles (2Y, 3B, 4G, 6R) and a table with 4 columns and 6 rows

- Help the child label the columns: Blue, Yellow, Red, and Green.
- Ask the child to select one of the color tiles and to place an X at the bottom of the appropriate column; have him or her continue with the remaining color tiles.
- Ask which color has the most, which has the fewest, and the difference between them.

At-Risk Students

materials: 21 pattern blocks (1T, 2Y, 3B, 4G, 5R, 6O)

- Have the child create a table with 6 columns and 6 rows; have him or her label the columns: Blue, Yellow, Red, Orange, Tan, and Green.
- Ask the child to select one of the pattern blocks and to place an X at the bottom of the appropriate column; have him or her continue with the remaining pattern blocks.
- Ask which color has the most, which has the fewest, and the difference between them.

Advanced Learners

materials: pair of dice

- Ask the child to create a table with 6 rows and 11 columns; ask him or her to label the columns 2–12, and at the top, add the title: Possible Dice Rolls.
- Ask the child to find and fill in all the ways (36) of finding each sum with two dice; each column should include all the orders. (Example for 7: 1 + 6, 2 + 5, 3 + 4, 4 + 3, 5 + 2, 6 + 1)
- Ask which sum has the most ways (7) and which has the fewest. (2 and 12, 1 each)

- **Conducting a Probability Experiment**
- **Identifying the Mode and Range of a Set of Data**

English Learners

- Write the following numbers on the board: 15, 10, 6, 9, 7, 12, 12, 8.
- **range:** the difference between the least and the greatest number in a set
- Explain that *range,* has many meanings in English, but when used for a set of numbers it means the difference between the greatest and least.
- Ask the child to identify the largest number on the board; ask him or her to identify the smallest number; if necessary, have the child order the numbers from least to greatest.
- Draw a line segment on the board; at the left endpoint write 6 and at the right endpoint write 15; below this write the subtraction number sentence 15 – 6 = .
- Ask the child to say the range (or difference); point to the line while mentioning that all of the other numbers in the set fall between these two numbers.
- Ask the child to use a complete sentence to explain *range.*

Spanish Cognate: range rango

Inclusion Students

The child will identify the mode and range of a set of numbers.

materials: 12 index cards each with one of the following numbers: 2, 6, 7, 7, 7, 10, 11, 14, 15, 15, 16, 20

- Shuffle the cards and ask the child to arrange them in order from least to greatest; explain that if there is more than one of the same number, place the cards next to each other in the row.
- Ask which number there are the most of (the number 7), and review that this is called the *mode.*
- Ask the child to identify the least and greatest numbers; help him or her find the range.

At-Risk Students

The child will identify the mode and range of a set of numbers.

materials: three dice

- Ask the child to roll the dice and record the sum on a piece of paper.
- Have him or her continue 10–15 more times.
- Ask which sum was rolled most often and what this number is called. (mode)
- Ask the child to identify the least and greatest numbers (ordering the numbers, if necessary); ask him or her to find the range.

Advanced Learners

Journal Writing (Overview 14)

- Have the child write a letter to a first-grade student and write about 3 things he or she learned in math the first-grade student will like.

• Using a Calculator to Add, Subtract, Multiply, and Skip Count

English Learners

materials: calculator

- **calculator:** a small tool with a keyboard and a display used to make mathematical calculations
- Explain that calculators are valuable and convenient tools used in mathematics, science, and business because of their speed and accuracy; a calculator should not, however, replace the person learning the skills.
- Review the parts of the calculator—the on/off switch, the visual display, the number keys, the keys for the functions the child is familiar with (+, –, ×, =), and the clear key.
- Review how it is important to clear the calculator between calculations, to be sure to get the correct answers to problems.
- Complete a few calculations with the child, describing the keystrokes as needed.
- Ask the child to use a complete sentence to describe a *calculator.*

Spanish Cognate: calculator calculadora

Inclusion Students

materials: calculator

- Review the parts of the calculator—the on/off switch, the visual display, the number keys, the keys for the functions the child is familiar with (+, –, ×, =), and the clear key.
- Review how it is important to clear the calculator between calculations, to be sure to get the correct answers to problems.
- Write the following problems on a piece of paper and ask the child to write the answer to each: 8 + 3 = , 9 – 6 = , 10 + 10 = , 10 × 2 = , 2 + 2 + 2 = , 2 × 3 = .
- Ask the child to use the calculator to check each answer.

At-Risk Students

materials: calculator

- Review the parts of the calculator—the on/off switch, the visual display, the number keys, the keys for the functions the child is familiar with (+, –, ×, =), and the clear key.
- Review how it is important to clear the calculator between calculations, to be sure to get the correct answers to problems.
- Write the following problems on a piece of paper and ask the child to write the answer to each: 8 + 3 = , 9 – 6 = , 10 + 10 = , 10 × 2 = , 2 + 2 + 2 = , 2 × 3 = , 60 – 20 = , 20 + 40 = , 4 + 3 + 2 + 1 = , 10 – 5 + 2 = .
- Ask the child to use the calculator to check each answer.

Advanced Learners

- Ask the child to write a paragraph explaining why it is important to be able to do mathematical problems without the aid of a calculator.

Guide to Differentiated Instruction

- **Choosing an Appropriate Method for Finding the Answer to a Problem**
- **Using a Calculator to Compare Data**

English Learners

- On the board write: Blue Team 3, Green Team 1.
- **score:** the action of getting points in a game; the number of points obtained
- Explain that the word *score* has several meanings in English; when used for a game, one meaning is the action of getting points, such as: He scored a goal!
- Another meaning is the number of points collected throughout the game by the individual players and by the teams; the team scores are used to decide the game winner, such as: Blue Team 3, Green Team 1.
- Ask the child if he or she plays or is interested in any sport; discuss how points are scored in that sport.
- Ask the child to use a complete sentence to explain *score.* (Either meaning is acceptable.)

Inclusion Students

The child will use a calculator to compare data.

materials: calculator and 6 index cards; write one of the following values on each card: 18, 21, 28, 35, 36, 44

- Shuffle the cards and give them to the child; explain that you will pretend that the cards show the scores from a game you have played.
- Ask the child to sort the cards from lowest to highest; ask him or her to identify the lowest and highest cards and to use the calculator to find the range. (26)
- Have him or her use the calculator to find the sum of all the scores. (182)
- Suggest that when adding many numbers on a calculator, it is a good idea to repeat the calculation, in case a number was entered incorrectly; if the sums do not match, do the addition a third time.
- Ask the child to find the sum of all of the even numbers (126), all the odd numbers (56), and the difference between them. (70)

At-Risk Students

The child will use a calculator to compare data.

materials: calculator and 7 index cards; write one of the following values on each card: 128, 135, 136, 144, 165, 165, 174

- Shuffle the cards and give them to the child; explain that you will pretend that the cards show your bowling scores.
- Have the child sort the cards from lowest to highest; ask him or her to identify the lowest and highest cards and to use the calculator to find the range. (46)
- Ask the child to identify the median (144) and mode. (165)
- Have him or her find the sum of the highest three scores. (504)
- Ask the child to find the sum of all the even numbers (582), all the odd numbers (465), and the difference between them. (117)

Guide to Differentiated Instruction

- **Identifying Acute and Obtuse Angles**
- **Describing and Classifying Plane Figures**
- **Identifying Pentagons**

English Learners

- On the board draw three angles—right, acute, and obtuse; show the angle marker on the right angle.
- **acute angle:** an angle that is smaller than a right angle
- **obtuse angle:** an angle that is larger than a right angle
- Review that a right angle has a square corner, which is sometimes shown with a small square.
- Remind the child of the differences between a right angle and an acute angle, where the lines are closer together, and an obtuse angle, where the lines are farther apart.
- Ask the child to use a complete sentence to describe *acute* and *obtuse angles.*

Spanish Cognates:	acute angle	ángulo agudo
	obtuse angle	ángulo obtuso

Inclusion Students

The child will identify acute, right, and obtuse angles.

materials: 1 of each pattern block

- Ask the child to identify the numbers and types of angles on each piece. (Example: parallelogram—2 acute and 2 obtuse angles)

At-Risk Students

The child will identify acute, right, and obtuse angles.

materials: 1 of each pattern block, plus the square, parallelogram, and three sizes of right triangles from the tangram set

- Ask the child to identify which pieces have the following:

1. only obtuse angles (hexagon)
2. any right angles (squares and right triangles); have him or her identify the other angles
3. both acute and obtuse angles (parallelograms and trapezoid); have him or her identify them
4. only acute angles (acute triangle)
5. only right angles (squares)

Advanced Learners

- Ask the child to draw 3 different pentagons; ask him or her to include an acute angle, a right angle, and an obtuse angle in at least one of the pentagons.
- Have him or her draw 3 different hexagons; ask the child to include an acute angle, a right angle, and an obtuse angle in at least one of the hexagons.
- Depending on the figures drawn, show that some angles may be outside (exterior) angles.
- Have the child use a blue crayon to trace the acute angles, a red crayon to trace the right angles, and a yellow crayon to trace the obtuse angles.